D0281379

The Beauty in the Beast

By the same author:

A Prickly Affair: The Charm of the Hedgehog

The Beauty in the Beast

*Britain's Favourite Creatures and
the People Who Love Them*

Hugh Warwick

**SIMON &
SCHUSTER**

London · New York · Sydney · Toronto · New Delhi

A CBS COMPANY

First published in Great Britain by Simon & Schuster UK Ltd, 2012
This paperback edition published by Simon & Schuster UK Ltd, 2013

A CBS COMPANY

1 3 5 7 9 10 8 6 4 2

Simon & Schuster UK Ltd
1st Floor
222 Gray's Inn Road
London WC1X 8HB

www.simonandschuster.co.uk

Simon & Schuster Australia, Sydney
Simon & Schuster India, Delhi

A CIP catalogue record for this book is available from the British Library.

ISBN: 978-0-85720-396-0
ISBN: 978-0-85720-397-7 (ebook)

Typeset in Fournier MT by Hewer Text UK Ltd, Edinburgh
Printed in the UK by CPI Group (UK) Ltd, Croydon, CR0 4YY

To the three most beautiful beasts:
Tristan and Matilda, and their mother, Zoë

Contents

Foreword

Hugh Warwick's book *The Beauty in the Beast* comes at an opportune moment. It is a gentle weapon of war against those who threaten the well-being and the very existence of our precious and entirely innocent wild animals. It is timely because we all now stand at a crossroads which will determine how the human race goes forward – either in harmony with the bountiful riches of life on this blue planet, or selfishly and ignorantly, plunging the world into a sterile abyss in which humans have obliterated the rest of life on Earth.

Books that encourage us to appreciate and love the natural world are more important than ever. We have become so far removed from the magic of Nature that we need strong reminders to reconnect us. The concerns may strike us when we take a moment to wonder what kind of a world our grandchildren will inherit. We may also begin to be concerned about ecology, the balance of populations and the survival of species; biodiversity has at least become a word we are all familiar with. But a true enlightenment only comes when we realise that our concerns must go much deeper than survival of species. It is when we realise that every single creature on Earth *matters* that we come up against the shocking discovery that the human race has veered horribly off the tracks.

Britain, in particular, prides itself as a nation of animal lovers, yet we have turned a blind eye to a mountain of cruelty and abuse for hundreds of years. In the present day, where the Internet enables us to see into every shady corner of human activity, there is no longer any excuse for allowing cruelty to continue – this applies to all creatures, whether human or not.

My own love of Nature has always been part of my make-up but it was a gradual growth of understanding of how cruel we really are to other species that led me to take up arms against the abuse of animals. Along with a consortium of animal-aware campaigns I have been working for the past few years on ridding the countryside of the inexcusable behaviours that are justified by 'tradition', or blinkered views of farming husbandry which place the value of a wild animal at zero. We who work in Animal Welfare are determined that wild animals and all creatures shall have a voice, in public affairs and eventually at government level, whereas at the present time they have absolutely no representation.

Outdated toxic views of the world lead to the blood-hunting of foxes, stags and hares, to badger-baiting, dog-fighting, and to an impending massacre of Britain's most ancient family-oriented species, the British badger, in the so-called 'badger cull'. There is no suggestion that this is a cull in the proper meaning of the word – for the health of the herd – it is simply a random slaughter of mostly healthy animals. All kinds of attempts are made to justify these tragic aberrations by a government that has lost all touch with the real needs and wishes of the residents of these islands, human and non-human. It is our job to restore decency and sanity to the acts of our species and our nation for the good of those who are at present abused, for the good of the planet and, in the end, for our own welfare too. The world will be a destitute place when all that is left is a landscape overpopulated with humans and devoid of any other life.

Hugh writes not just about the power of compassion – of people who do not want to see wildlife killed – but also about the science, which roundly condemns this behaviour, and ethically why it is simply wrong. Hard decisions have to be taken as we try to balance the immediate perceived needs of humans with the last remnants of our natural world. But the evidence is there for all to see: that no good can come of the killing.

Any hope for a decent future depends on us acting in harmony with the life around us, not in conflict with it.

This is why *The Beauty in the Beast* is an important book. Gently wise, the facts are delightfully delivered with a good dose of humour. Warwick gives us every possible reason to fall in love all over again with the natural world; it is a love which, in the coming crucial months and years, will inspire us to fight for a compassionate world.

Brian May, January 2013

Introduction

With a noise like a dentist's drill being operated by a swarm of angry mosquitoes, the needle began to stab repeatedly into the flesh above my left ankle. Not for the first time that evening, I seriously wondered what on earth I was doing. Forty-three is too old to get a tattoo, I had been told. Yet there I was, on a small stage in an art gallery, in front of a crowd of people drinking beer, getting my first, and last, tattoo.

I was taking part in a project called ExtInked. Artist Jai Redman, a member of Manchester-based creative practice Ultimate Holding Company, had drawn pictures of 100 species from the UK's Biodiversity Action Plan and then set about trying to find 100 volunteers willing to become ambassadors, displaying their commitment to the animals, plants and fungi with a permanent tattoo.

Due to the minor celebrity accorded by the publication of my book *A Prickly Affair* the previous year, I was asked to be the very first. So there I was, feeling more than slightly out of place. I have been studying, rescuing and worrying about hedgehogs for twenty-five years. I am an ecologist and a writer. What was I doing here, undergoing an initiation into the wild world of be-inked people, with my trouser leg rolled up like a novice Mason?

Most of the observers and participants were well tattooed already, and the guys with the needle-guns, from one of the country's foremost tattooing teams, Ink vs Steel, were highly decorated. Does everyone have a tattoo these days? Was I in danger of becoming normal?

The pain. I had been expecting pain, lots of it. But it was rather innocuous. There was definitely something happening, it was not comfortable, but not what I would call pain. What was odd, though, was the strange taste, rather as if the ink had made its way to the back of my mouth.

Actually, what was odd was that I was having a tattoo in front of a crowd of strangers. As I looked around I began to wonder whether the other people awaiting their turn on the stage and under the needle were fellow travellers. Was this a room of people like myself? Perhaps it was the endorphins kicking in, but I began to imagine spending time with each of them as they tried to persuade me of the delights of their particular species. Could I get as excited by the shrill carder bee as I did by the hedgehog?

And then it was over; my artist wiped the last spots of blood and stray ink from my leg and applied a layer of nappy cream before wrapping my leg in clingfilm.

The next morning I woke slowly, alone, in a friend's flat and ambled to the shower. As I washed I suddenly worried that I might smudge the rather cute hedgehog that was now on my leg.

Tattooing takes time, and ExtInked was to continue over the weekend. I headed back to the gallery in the hope of meeting more obsessives. What was it that motivated people to become indelibly linked and incredibly inked with a particular species?

As wonderful as the other ambassadors I met were, I was surprised to find none who had come to the project on the back of a lifetime's work. So was I unusual? Was I alone in dedicating so much time to just the one species? I could feel an idea bubbling. I wanted to find other people who shared my fascination in wildlife but had taken, possibly, one or two steps outside the bounds of what most people considered normal. Was there an otter version of me, wading through streams and getting nose-to-nose with Tarka? How about moths, or robins or toads? I knew people who were fascinated by particular species. But could I find those who were a little more than merely interested? People who found beauty in the most unlikely of beasts; people who proselytised on behalf of their chosen animals. In other words, advocates and ambassadors for Britain's wildlife.

I preach Hedgehog, frequently. And the story of my tattoo has become a staple in the lectures I give. Not just because this emblem of my midlife crisis is an easy way of breaking the ice, but because it posed an important question. What was the hedgehog doing on the Biodiversity Action Plan in the first place? This is the most common or garden of animals, one that everybody can recognise. Is it really threatened?

The depressing answer is yes. I have been helping the British Hedgehog Preservation Society and the People's Trust for Endangered Species to understand the full extent of the problem; one that I had originally noted in purely anecdotal form as a result of questions I was asked at talks I gave to the Women's Institute. Every time, there was the same question: 'Why are there fewer hedgehogs?' And as much as I hoped this was the result of rose-tinted nostalgia for better times gone by, the questions identified a trend that is all too real.

The best data we have now shows a very clear – and statistically significant – drop of 25 per cent in Britain's hedgehog population in the last ten years. What is more worrying, though, is the evidence of a 95 per cent decline in the last sixty years.

All over the country there are shocking declines in the numbers of many other species, declines that often go unnoticed because the species concerned are not as charismatic and loved as the hedgehog.

That was part of the delight of ExtInked; it was bringing the challenges facing some of the far less well-known species out into the daylight for the first time. And while there was expected to be a clamour for the obvious, like the otter, the organisers were thrilled that obscure fungi and insects were equally in demand – even if this passion was sometimes as much for the art as for the organism it depicted.

My own passion, for hedgehogs, had led to me making the bold claim, towards the end of *A Prickly Affair*, that the hedgehog was the most important species on the planet; that in its grubby little paws was held the key to the salvation of humanity.

I knew I was asking for trouble. Admittedly my criteria were peculiar. The importance of the hedgehog comes not from its economic impact. It is not a voracious destroyer of crops nor is it a pollinator of plants or a spreader of disease.

For me the importance of the hedgehog comes from its ability to act as a gatekeeper to the natural world. The use of the so-called charismatic mega-fauna, like lions, whales and elephants, to sell a love of the natural world – as is done by most of the wildlife and conservation organisations – is all well and good, up to a point. But it keeps us at a distance. They are like the celebrities of the animal world: we can look at them, but we cannot touch. True love is felt for the boy or girl next door, not the pop star or actor. And the hedgehog is the animal equivalent of the girl or boy next door.

So, as I argued, we need to change our perspective, get nose-to-nose with the hedgehog and let the love in. Because, as biologist Stephen Jay Gould said, 'We will not fight to save what we do not love'. This is why the hedgehog is so important.

But there were other people beginning to take issue with my claim that the qualities of the hedgehog were unique. And gradually I realised that they might have a point. While the hedgehog met my needs, other animals would do it for other people – and there was a bigger issue, about which I felt passionately, and that was finding ways for people to fall in love with nature. And that issue was not just about increasing the membership of the British Hedgehog Preservation Society; it was a real desire for people to take their passion further.

Concentrate on one component of this wonderful world and allow it to suck you in, deep. This is not new; William Blake had it in mind:

To see a world in a grain of sand
And a heaven in a wild flower,
Hold infinity in the palm of your hand
And eternity in an hour.

My attention is forever being drawn. I cannot walk by a lavender bush without gently gripping the fragrant flowers or pass a clump of jasmine without breathing unusually deeply. I get moved by goldfinches, our urban skylarks, song thrushes and swifts, and find fur irresistible. When the moon is behaving outrageously with a display of golden fecundity, I call my friends and get them to look up. I am a bit of a promiscuous sensualist, but I always come back to my first love, the hedgehog.

But there must be people who are equally obsessed about water voles, badgers, foxes and owls as I am about hedgehogs. By spending time with other ambassadors I would uncover other ways of drawing people into a love of nature. And there must be some animals that are simply better suited to this role than the hedgehog. I started making a list of people and animals. Who would be able to win me over to their totem?

My first list was intimidating. Even the shortlist held nearly thirty species.

I had to cut it down, but the process of discarding such potential was painful. Sometimes I was helped: crows, for example, were an obvious choice of gatekeeper, but they were also very well covered by modern nature writing. Similarly butterflies. I had wanted to write about peregrine falcons, but there was no way I could create anything to compare with my favourite piece of nature writing, J.A. Baker's *The Peregrine*.

I went some way to having a swan chapter. There is a passionate swan advocate up in York. Dan Sidley set up the Yorkshire Swan Rescue Hospital and he invited me up to a fund-raising ball in the Railway Museum. Somehow I ended up as the auctioneer. But Dan's ambition to become a councillor and then MP kept him too busy to extol the virtues of this majestic bird.

In the end, moles and hares fell off simply because I wanted to resist the temptation to inhabit my list entirely with mammals. And as hard as I tried to find an advocate of the stone-boring piddock, no one came forward to make their case. Which is a shame. It is an amazing mollusc.

Surrounded by the enormous range of species on offer, I felt a little like a hedgehog snuffling through long grass. It was almost overwhelming, but I found myself being drawn towards tasty morsels, and I have snaffled them. What follows is far from a comprehensive guide to the animals of Britain; it is an eccentric, eclectic and entirely subjective wallow in the wonders of the country's wildlife. And none of it would have been possible but for the equally eccentric individuals who gave up their precious time to try to seduce me away from my hedgehogs.

But they had an incentive. As I formulated my plan, more than one friend saw straight into my subconscious and declared that all I was doing was trying to justify another tattoo. Which is strange; to start with, that was definitely not my intention, but a seed had been sown that, by the time I embarked on a year of travels around the wildlife obsessives of Britain, had grown into a great beast of an idea. More than that, it seemed like a logical conclusion to my quest. Not only had my ambassadors to convince me of the beauty locked away within their chosen beasts, but to persuade me to such an extent that I would become permanently tattooed with the creature in question.

And this would most definitely mark the end of my midlife crisis. No more tattoos.

1. SOLITARY BEES

'Furry, like an impossibly light mouse'

Sit still for a moment and watch the bees. Wait for a beautiful early-summer day. Find a spot near a flowering apple tree or privet hedge or a field of flowers, and then watch. Really watch. Don't just look, see. There is a host of small insects, which at first glance may be dismissed as 'flies' but are, on closer inspection, bees.

My eyes were opened to this world of bees by a man who, at times, can barely see. And now they have been opened, I find myself easily distracted by a buzzing bush, losing great chunks of time as I gaze at the bees.

I first met Ivan Wright, my bee gatekeeper, at an allotment open day in Oxford. He was there on a mission to persuade the passing masses that bees are good; and not just the honeybees or bumblebees with which we are all familiar, but the unsung heroes of our countryside, so-called solitary bees.

The bulk of a bumble is helpful. Ivan's opening gambit was to hold his hand out to my daughter and ask if she wanted to stroke the large bumblebee that sat there. She extended a tentative finger and touched the insect. As I watched, she stroked the bee briefly, before seeming to get self-conscious and backing away.

Then it was my turn. Ivan obviously sensed some wariness.

'This is a male buff-tailed bumblebee,' he assured me. 'Male bees cannot sting.' And as the gently vibrating bee walked onto my finger he explained how males do not possess the modified ovipositor that makes the females quite so spiky. The bee was so furry, like an impossibly light mouse.

There was a glint in Ivan's eyes; was he sensing the possibility of another convert?

When we next met it was over a pint in a local pub and Ivan was keen to impress upon me the importance of bees. Yes, he was passionate about honey and bumble. But it was when we came to solitary bees that the evening started to get really interesting.

To start with there is the diversity. In the UK we have just the one species of honeybee, twenty or so species of bumblebee and around 220 species of – largely overlooked – solitary bee such as the mason and the leafcutter bees.

And then when you take a closer look, the extraordinary nature of these little insects becomes clear. For these bees are as varied as jewels: some are black, some metallic green, some red and black, others grey. Some are hairy, some are not. Some look like miniature versions of the more familiar members of the bee family, others most certainly do not; indeed, some look almost ant-like, which should not be a surprise, since ants, wasps and bees all belong to the same taxonomic grouping, the aculeate hymenoptera: stinging (aculeate), fine-membraned (hymen), winged (optera) insects.

The solitary bees are, he argued, far more efficient pollinators than honey- or bumblebees.

'My apple tree, for example,' he said. 'If you pay attention you will see that the pollination is not carried out by the honeybees in the area, but by just a few dozen red mason bees. You would need hundreds of honeybees to get the same effect.' The red mason bees have very hairy bellies, specifically adapted for collecting loose pollen: the way the bee moves causes it to scrape the pollen-coated abdomen over the stigma of the apple flowers, increasing the chances of successful sex.

Then Ivan reeled me in. 'Would you like to come and join me doing some fieldwork?' he asked nonchalantly. 'I could do with some help.'

So that's how I ended up carrying a large sack of yellow frisbees up a wonderfully wooded hill to the east of Oxford, very early one Sunday morning in June.

Ivan has experimented with different objects for this task and has concluded that nothing beats yellow frisbees (or should that be frisBees?). The colour attracts the insects and, when the frisbee is upside down, it creates a shallow dish that, filled with water, becomes a puddle for passing insects.

It seems unfortunate that Ivan ends up killing quite so many of the animals he loves. They approach the yellow disc, which is so like a large flower to their multi-faceted eyes; when they land, they get trapped. He adds a little soap to the water to reduce the surface tension and ensure that even the smallest insects sink. But surprisingly they don't drown. Only those insects that need microscope examination have to be killed.

I questioned Ivan's apparent callous disregard for his bees, but his response was calm.

'The only way to identify the species of many of these bees is under the microscope. Anyway, there is no way I can kill as many as the birds eat, and there is always the argument that they are dying for the good of their species, though obviously they have not had much say in that state of affairs.'

For this part of Ivan's studies, we laid the frisbees in lines known as transects, from the top of Shotover Hill out into the surrounding countryside. The aim, Ivan explained, was to find out how far the bees travelled in search of food.

The question of the bees' dining habits is not just academic. Ivan is hoping that the bees will help to protect Shotover. Because underlying Ivan's obsession with the minute beauty of bees is a passion for place.

Ivan has lived on the borders of Shotover for over twenty years. Shotover Hill is a remnant of the large medieval royal forest of Shotover that almost encircled medieval Oxford. There are steep slopes, ancient oaks and well-worn paths. It is pleasingly wild, big enough to lose oneself in but not so big as to get lost.

The steepness of the hill has held development at bay for generations, but as the demand for land has increased, so has the potential threat to Shotover from developers and planners.

Fifteen years ago Ivan started having great difficulty with his eyesight. Fortunately, treatment halted the deterioration of his cornea, but he was forced into early retirement from his career as a microclimatologist. Now, for a limited number of hours every day, Ivan wears special lenses that allow him to see well enough to drive; but as soon as the lenses come out, 'I'm all white stick and helpless,' he explained.

Early retirement allowed Ivan to concentrate on trying to find ways of promoting the importance of Shotover for wild-life. The best bet was to increase the protection granted to the area by central government. And the best way to do that was to find a rare species or habitat that got people to sit up and take note.

Ivan's wife Jacqueline is a botanist, specialising in mosses and liverworts. So he contemplated joining her, focusing on these plants. But then came a conversation that has helped shape his life.

'Steve Gregory, an old friend and an excellent naturalist, just made a passing comment: "Someone should be studying the bees and wasps up there". So I thought I would give it a go.'

For most of us there really are just two sorts of bee: honey and bumble. But as soon as you start to look at the diversity of solitary bees, you quickly run into trouble. 'The easy ones are easy to identify, but the hard ones; they can be a night-mare,' Ivan said.

To tease apart the species of bee requires a 'key'. Keys are magical devices employed throughout science in order to unlock knowledge. They take the form of a series of ques-tions, each offering multiple answers; the selected answer determines the next question – and so on, until you are able, in this case, to identify the species.

For example, here is the beginning of a key to identifying the main taxonomic groups of insects:

| 1a | Insect has conspicuous wings | 2 |
| 1b | Wings inconspicuous or absent | 11 |

| 2a | One pair of wings | 3 |
| 2b | Two pairs of wings | 4 |

| 3 | One pair of wings | Diptera |

| 4a | The two pairs of wings are unlike in structure, first pair thicker than second | 5 |
| 4b | The two pairs of wings are similar in structure, about the same thickness | 8 |

| 5a | First pair of wings hard and shell-like | Coleoptera |
| 5b | Some part of first pair of wings leathery | 6 |

And so on, until you have sorted your insects into the first level of the taxonomic tree. Ivan's key starts with far more detail and reveals the process of unpicking the species to be as much art as science. Just plucking a paragraph from Ivan's 'bible', a delightfully battered notebook, made me realise how little I knew:

50 (The females in this couplet are difficult to distinguish). Middle furrow of propodeum boat-shaped, widest at middle or rear, more distinct from front furrow; mesonotum with pits sparser, rather shiny between them; (scutellum usually without yellow spot); *Crossocerus tarsatus*.

He still has the book containing the key that allowed him to enter this new world. When it arrived, Ivan thought he would give it a go. Not expecting a great deal he took his net up to the hill, caught some bees and returned to one of Jacqueline's microscopes. He found he could do it; he could work out exactly what species he was looking at.

But wait, this is a man who suffers from not-inconsiderable visual impairment. And a great deal of bee work is done under the microscope, as this is the only way to separate out many individual species. Amazingly, Ivan can see clearly through a microscope, because the light passes through a very small portion of his corrupted cornea. In normal vision, however, light falls across all of the now rather rumpled surface, making detailed observation difficult.

After about eighteen months of collecting, analysing and recording, Ivan decided to take his results to Steve Gregory, to see if he was doing it properly. 'I asked him about my results. He just looked at me and said, "How should I know; you're the expert now!"'

So often in life it is when someone else shows faith in your abilities that you develop the confidence to tackle seemingly impossible tasks.

So, Ivan had learned a new skill, but to what end? There are entomological 'stamp collectors' out there; people who collect insect specimens just for the sake of collecting. They have musty drawers stuffed with pinned specimens; what excites them is the goal of completing their collection. But that was not why Ivan started.

The data Ivan collected are beautiful and fascinating. He compared what has been seen on Shotover in the past with what is there now. For example, sixty-two species of aculeate hymenoptera that were recorded before 1939 have vanished.

What is just as remarkable is the discovery that there are sixty-eight *new* species now resident, which used not to live on Shotover.

So why such a change? When Ivan went back through the historical documents he found that not only had the bees changed, so had the habitat, from pasture and heath to woodland. Careful analysis showed that the nesting requirements of the 'formerly recorded' and 'recently added' insect species differ, and those differences were consistent with the changes in the insect population.

The data revealed Shotover as a 'hotspot' for bees in Oxfordshire, one of the best sites in the county, with ninety-nine bee species found in the area by Ivan between 2000 and 2004. So he was able to argue for improved protection for Shotover, and enhanced its status as a Site of Special Scientific Interest.

Solitary bees have helped to protect the hill, and now the reason why we were laying frisbees out in transects became clearer. The bees do not spend their entire lives up on the hill; it does not provide the diversity of plants they need to survive. So the bees need to commute to feed on nectar and pollen from a variety of flowers; the nectar provides sugar for energy and honey, while pollen provides protein.

And if the land around Shotover is desertified by oil-seed rape and concrete, the bees will disappear, as they will have no chance to feed themselves or their subterranean grubs. So Ivan wants to expand the area that is safeguarded to include some of the agricultural land that, when managed sensitively, can generate rich sources of bee food. The lines of frisbees radiate out from the top of the hill into the surrounding farmland. Proving that the bees commute along the lines formed by the frisbees will give Ivan the evidence he needs to argue his case.

Modern farming has seen a very distinct link between the increase in agrochemical use and a decrease in the diversity of flower species. The statistics are depressing: according to research done by R.M. Fuller in 1987, this country has lost 97 per cent of its wildflower meadows since the end of the Second World War. These were the meadows that inspired poetry with their rich diversity of flowers changing with the slow waves of the seasons; they were hotspots that fed the national pollinating effort.

Over sixty frisbees later we returned to Ivan's bungalow beside the wood. We would not need to collect the discs until early evening.

As we settled down with a mug of tea Ivan told me to look at the small plastic tub just behind my head. In it there were three small cardboard tubes tied together. He reached over and lifted them up, cooing gently as if coaxing a nervous child out from behind a sofa. 'Come on, now, Harriet, no need to be nervous.' And, as if on command, there appeared the beginning of something quite enormous; it – or rather 'she', as I was corrected – kept growing. What emerged was a queen hornet that looked like a giant wasp, over three centimetres long. Ivan had found her, apparently lifeless, while walking through the woods. So he picked her up and took her home, only to discover she was still very much alive, but unable to fly.

She had lived with Ivan as a houseguest for three weeks, and if she started flying, said Ivan, she would be asked to leave. As she crawled up Ivan's sleeve, my flesh crept. She was beautiful, but she did not leave me feeling relaxed.

The tea break was a chance to learn more about solitary bees. As their name suggests, they do not form the great communities of their more famous cousins. A honeybee hive may contain over 20,000 individuals, regimented into different roles: workers (females that do all the work), drones (males that mate and get fed) and the queen (the egg layer).

Even the bumblebee can create a colony of up to 200. But the solitary bee is just that: solitary. Each female is fertile and constructs her own nest. She does not need to expend energy on honey and wax; the nest is just a series of cells in which she lays her eggs with a store of nectar and pollen for when they hatch.

Our break was over and we headed back out, this time looking more like traditional entomologists: we had nets and sun hats.

Shotover is a wonderfully varied place. The steep slopes are studded with old trees hung with rope-swings that would leave any health-and-safety official reaching for their clipboards. In fact, Ivan took me to one of the swings we regularly went to with the children, and told me to look carefully. Just beside the branch supporting the rope there was a stump and, as my eyes acclimatised, I began to see what was making Ivan grin. In the stump was a hole, busy with bees.

They were honeybees. They had been there for three years, probably escapees from a hive. And they seemed to be thriving, unlike many cultivated bees. Around the world there have been sudden, catastrophic bee deaths, a phenomenon now known as Colony Collapse Disorder. Whether this is the result of agrochemicals, climate change, varroa-mite infestations, loss of food plants, stress or – as is probably the case – a combination of these and other factors, we do not know.

As we watched the free bees, Ivan introduced me to some honeybee animal-rights issues. Amazingly, one third of all honeybees living in the US spend their entire lives in articulated trucks, being shunted from state to state, pollinating great orchards of fruit and nut trees. 'The worst are the almond orchards,' he explained. 'Honeybees don't like almond nectar. So to make sure that the bees do their job, the ground is sprayed with herbicide, killing everything else. There is nothing else for the bees to eat, so they stay with the almonds. These are some of the grumpiest bees you will ever meet.'

The consequences of further colony collapses, among other species of bee, would be disastrous, as it would remove vital plant pollinators. Ivan just hopes that whatever is doing for the honeybees leaves the other bees alone.

Other bee species are already being used by industrial food producers. The largest salad producers in the UK, for instance, are the monumental 'Thanet Earth' greenhouses in Kent, where bumblebees are used for pollinating – thousands and thousands of great big *bombus* flying around the tomatoes and peppers.

We left the honeybees buzzing around the stump and headed on to see what was really exciting Ivan – an absolute wonder of wildlife management and the best bee-conservation work he has been involved with. I was pulled up short as we stepped from behind a large oak; about 300 square metres of land had been stripped. It looked as if a bulldozer had been through there. It had.

'You see, we often base our ideas of the value of landscape on a very simple aesthetic,' Ivan explained. 'I helped to raise money from some developers to do this. Last year we stripped this land, took off the topsoil and the monoculture of bracken. Now we can watch the regeneration and already it is fantastic.'

Even in this short space of time small cones of soil had sprung up across the area, possible evidence of tawny mining bees. These bees burrow down into the earth and often cause consternation as they emerge from lawns in the spring. I had come across them before, oblivious to their solitary nature, when a more alert friend spotted one flying low over the playground of an Oxford primary school.

Something was wrong. It took me a moment to work it out: the school had put down fake grass around the climbing frames, and the bee was quartering the surface, searching for somewhere to dig a nest. But every time it reached the surface it would stay for a few seconds before moving on. There was no way through the green plastic to the soil.

Next to the ecological flaying of Shotover a great bite had been taken out of the hill; again, the work of Ivan's JCB. This was our target: a two-metre-high cliff with a three-metre sandy 'beach' at its base. The sandy wall of the cliff was covered in hundreds of small holes, as if pockmarked by a machine gun, each the nest of a solitary bee or a wasp.

To begin with Ivan just stood there, seeming to ignore the buzzing of insects. But then I saw him strike. His right arm flicked out, propelling the net at an extraordinary speed, before swiftly twisting it in a complex loop and returning it to his side with a slightly indignant insect inside. I was most impressed.

'It's the jizz,' he announced enigmatically, as he brought the net closer for inspection.

Jizz is a bird-watching term. With the benefit of experience, the identification of a species can become subconscious, based on just a fleeting glimpse of a bird or, in this case, an insect. It would be difficult to say exactly what features had allowed you to identify the species; it just 'felt' right. So what had Ivan caught?

As Ivan had told me earlier, the easy ones are easy, and this one was: *Osmia rufa*, better known as the red mason bee, the species that pollinated Ivan's apples. And as Ivan knew what it was there was no need to pop it into the pot of alcohol, so off it flew. 'Amazing things these mason bees,' Ivan said. 'The female makes a series of cells, separated by walls of mud, inside holes like the ones in front of us. She has two "horns" on her head, which she uses to mould the mud. In each cell she places a small pile of pollen, then lays an egg on top, before capping the cell with more mud. And she can determine the gender of her offspring. Females at the back, males at the front. After hatching, the grubs eat the pollen and then pupate, transforming from grub to bee in a sleeping bag of silk over winter. Here we are in early summer and the cycle begins again.'

Back to business. I was beginning to be able to ignore the flies and beetles. And I was beginning to *see* the bees, a facility that has stayed with me. Now, as I walk by my garden pond, I find myself noticing the differences between the insects as they fly. I am not expert enough to be able to identify individual species yet, but the ability to spot basic differences is a good place to start.

So now to catch a bee. I play – or rather used to play – a lot of squash. And I like to think I have a sharp eye. So why was I now missing every bee I swung for? Ivan was scoring with nearly every stroke, while I was taking great chunks out of the sandy wall, filling my net with dust. And when I did catch one it escaped as I tried to twist the net in order to contain it.

Ivan consoled me with the information that my crafty escapee was not a bee at all, but a very beautiful ruby-tailed wasp.

And then, just to rub it in, Ivan swiped and caught one of the same species. 'Probably the one you missed,' he said, as he carefully extracted the remarkable animal. The tail was red and the body seemed to sparkle, iridescent, in the sun. From one angle it was blue, another green, though it was so jewel-like that it would be better to say sapphire and emerald. There is a knack to holding such a small insect: squeezing firmly without causing an explosion of intestines is a skill few will have mastered.

Such a beautiful animal, with such a dark heart (if you will forgive a little anthropomorphism), the ruby-tailed wasp is solitary, and is what is known as a cuckoo wasp. Just as with the avian cuckoo, this specimen was looking for an unguarded nest to deposit her eggs. She would then depart, leaving the unwitting host to nurture the alien offspring. When her larvae emerged, they would eat both the bee larvae and the pollen store.

There were plenty of other opportunists out that day — both bees and wasps parasitise. But wasps have the upper hand as they are carnivorous, and some are hunters. *Oxybelus uniglumis*, the common spiny digger wasp, not only stings its prey; it then carries it back to its hole impaled on its sting.

It is these parasitic wasps that have led to the evolution of the bee's sting. The sting is not there to torment us; in fact, it is very poorly adapted to loose-skinned mammals like humans. The sting is designed to penetrate between the plates of the exoskeleton of a marauding cuckoo bee or wasp. The sting of the honeybee is barbed and gets caught in our skin, causing it to be left behind along with a good portion of the bee's intestines. Which in turn leads to the demise of the unfortunate bee.

Bumblebees do not have a very barbed sting and so can extract it and use it repeatedly. Solitary bees, on the other hand, while having a sting, very rarely use it on us, as they have no big colony to protect. I noticed that one of the bees Ivan held was more intent on trying to bite him than sting him, a peculiar strategy that caused him no more than a mild tickle.

Ivan was particularly pleased to be finding parasitic bees and wasps, as their presence indicated the presence of the host species, killing two birds with one stone.

The sun was warm, there was the gentlest of breezes, and I could hear a midday chorus of birds: the whitethroat singing its heavily punctuated song, and then a blackcap's song, which was similar, but with more commas and fewer full stops.

Net poised, I slipped into the sort of delightful, meditative mind space that comes from complete concentration. There are spiritual practitioners who charge a small fortune to help people to this state, and here I was, with sun hat and net, drifting towards enlightenment. At Ivan's suggestion I tried my luck with a passing bumblebee. He thought it might help me get my eye in. But I missed and emerged from my trance long enough to ask an apparently peculiar question. How could I fail to catch an animal that cannot, according to physics, even fly? I had heard that the bulk of the bumblebee is too great to be elevated by its wings. Ivan explained that, based on earlier applications of physics, bumblebees indeed cannot fly. It was only relatively recently that scientists had begun to understand how the vortex created by the bee's wing beat enabled it to lift its considerable weight.

Satisfied, I returned to 'the zone'. And then I got one. I might have lacked Ivan's grace, but I did it. I left the bee extraction to Ivan, who glanced at my catch before dropping it into the pot for its final drink. 'A nomad,' he declared. 'One of the *Nomada*.'

As far as I was concerned this small bee looked like a small wasp. Studying bees seemed much more leisurely than studying hedgehogs. I was beginning to enjoy myself, and simultaneously feeling a little guilty. Would I really exchange the cold, wet nights for this? Ivan decided to twist the knife: 'You know, it is impossible to do this sort of work unless the weather is really nice. The cold and the rain and the night all drive the bees underground.'

But bee work is not all just grand flourishes of nets while meditating in the balmy warmth of the British summer. And eventually Ivan decided it was time to head home. There was deskwork to be done, and he had promised to involve me in this as well, despite my ignorance.

My job was simple, although with Harriet the Hornet peering down at me, not without an edge of danger. I had to separate the bees from the flies among the samples we had caught. Simple? Well, after some time I did get to the point where I was able to identify the flies (they have just one pair of wings). But in the time I had sorted one pot, Ivan had managed eight.

Leaning over my shoulder, he shifted a dish so that one particular insect came into view under the lens of the microscope I was using. '*Nomada goodeniana*; possibly the one you caught,' he declared. 'Now, this little wasp-like bee is another sneaky kleptoparasite. She will watch and wait while one of the ground-nesting bees, such as *Andrena haemorrhoa*, the early mining bee, lays her eggs and gathers nectar and pollen for the grubs when they emerge. The nomad can then be seen sniffing around the small holes that lead to the mining bee's underground egg chamber, and when she finds one that is well provisioned, she quickly lays her eggs. The mining bee continues to collect food for her brood, unwittingly ensuring the cuckoo bee's future.'

The future of all of the solitary bees in the UK is being helped by a network of people like Ivan: amateurs as experts. They feed data into local and regional recorders who then pass it up to the Biological Records Centre (BRC). The informal networks help to maintain a national overview of the state of many species. And it is not just allowing them to keep an eye on the bees, but also to try and identify warnings that will help our species as well.

Climate change can cause effects on wildlife before we even notice it. Subtle shifts in the breeding habits of insects and birds can reveal a great deal – a gradual move northwards of a species, for instance, to cope with increasing temperatures. For Ivan, his contact point for the BRC is the Bees, Wasps and Ants Recording Scheme. But because there are far fewer collectors of bee data than, for example, collectors of bird data, the maps produced are inevitably incomplete. In some places, for instance, it might look like there is a nationally important concentration of a bee species, but this may simply be because someone in that area happens to be looking for bees. 'This is one of the reasons why I am so keen to enthuse potential recruits to the bee cause,' said Ivan. 'We need people out there, all over the country, collecting data.'

The sun was beginning to hint at the sumptuous saturation of evening by the time we had finished, and we still had one job left to do: collect the frisbees. And, due to childcare complications, I now had an assistant in the form of my seven-year-old daughter, Matilda.

It was going to be quite a long walk, and on finding that the grass in the hay meadow came up to the top of her head, Matilda was having second thoughts. But when she saw the buttercup field, her face just lit up. The field had been impressive in the morning and was now glowing in the lowering sun. Mati looked like a little mermaid in a sea of yellow and gold. The buttercups and the next wave of flowers that would replace them; it was all hay-to-be, a sea of bee food.

At each stop, we strained the bees from the frisbee's water into jars of alcohol. Mati took on the role of flicker-in-chief. To clear the tea(bee)-strainer, a deft flick was required.

Every bee we found further from the top of the hill was evidence of the extent of their foraging range. In the frisbees were grey mining bees, brassy mining bees and the red mason bees, along with many other smaller animals that Ivan would need to identify back at home. There would be plenty for Mati to tell her friends the next morning at school.

She was obviously a bit young to take on board the importance of the work that Ivan does. But there is a terrifying statistic that may come to haunt her and her generation in years to come. Already over 70 per cent of the UK now has fewer bee species than it did in the 1980s. Even our gardens, once a bastion of diversity, are being colonised by larger blooms: double flowers that have no nectar or pollen, or are so floppy and unstable that bees cannot gain purchase to crawl in to feed.

The more I learn about the importance of bees, the more convinced I am that Ivan's passion is one of the most logical I have come across. He is delighted to share his passion – with me and Matilda, or at schools, fêtes and festivals. He wants others to be touched *by* bees as much as he wants them to touch the bees physically; he wants people to understand how integral bees are to so much of what we take for granted. But he also wants people to stop, and appreciate the wonderful complexity of the natural world.

When you see, as I did that day, a hairy-footed flower bee, flying with its tongue out, in anticipation of a tasty sample of nectar, it is hard not to be amused. Likewise when the male wool carder bee tackles a potential competitor over its jealously guarded patch of flowers. I was watching one chasing off far larger bumblebees, when Ivan told me of these diminutive warriors' fondness for flowers like lamb's ear, mullein and pelargonium. A male will patrol a patch in an attempt to lure in a female and is known to kill even honey-bees in its defence. The reason for his prowess in the martial arena is thanks to three spikes on the end of his abdomen, with which he can crush a rival.

So why the folksy name? Why not call it the 'bee-crusher bee', for instance? It turns out that the female is even more creative than her partner, scraping the hairs off hairy-leaved plants, and then, back at the nest, carding out the hair like wool to line the cells into which she then lays her eggs.

The study of these bees is a lifetime's work, and Ivan will never run out of things to learn. Nor, judging by the time I have spent with him, will he ever run out of enthusiasm – for Ivan still possesses a childlike glee when he works. Now, whether that is the result of those hours of meditating, net poised, waiting for bees, or whether to enjoy the meditating and the waiting you need to be a gleeful sort of person, I don't know. But Ivan has that sense of wonder; the same sense of wonder I feel when I see a hedgehog. This, I think, is one of the key elements of a successful relationship with another animal: the ability to retain that sense of delight.

Now comes that most difficult assessment. I am utterly seduced by Ivan's love of bees, and I was amazed by how rapidly I sank into a trancelike state while waiting with my net. I also like the fine-weather demands of bees. But while I can appreciate the importance of bees to civilisation, they lack an individual appeal. You can stroke the stingless males, yes, but they don't provide much opportunity for the sort of nose-to-nose proximity, nor the eye-to-eye flash of possible understanding that I experience with a hedgehog.

My time with Ivan left me tired but glowing: from the sun and the beauty on Shotover, but more than anything from the satisfaction of steeping myself in a completely new world. That night, as I bent down to untie my shoe-laces, I found that I had brought some of the wonderful day home with me, for the pale stitching of my boots had been coloured yellow by the pollen of uncountable buttercups.

2. BADGERS

*'Little teddy-bear ears atop a
black-and-white-striped face'*

If there was ever a species with its work cut out to convince
me of its loveliness, it has to be *Meles meles*, the Eurasian
badger. When one killed and ate Nigel during a radio-tracking
project in Devon, I was heartbroken. Nigel was my favourite
hedgehog and had let me into his life in a way that no other
animal ever had. One particular night out with him helped to
change my entire relationship with the natural world as we
gazed into each other's eyes, nose-to-nose on a country lane.

And I am hardly alone in having an – at least – ambivalent
attitude towards badgers. I might have been inflamed to brief
rage when one killed Nigel, but that hardly compares with
the community that has declared war on the badger. Many
farmers are not far from forming a mob to attack the castle of
legislation that protects this scourge of the countryside. Are
badgers not the cause of bovine tuberculosis? Have they not
stripped the countryside of hedgehogs and bumblebees?

And yet, despite losing some dear friends to badgers, I am
instinctively attracted to the species. We three have similar
habits, the badger, the hedgehog and me, of which nocturnal
rambling is key. We share a fondness for damp hedgerows,

too (though I draw the line at eating macro-invertebrates). Ambivalence is definitely the word.

The seduction began with a DVD: *Badger Watching with Gareth Morgan, (the Mid-Wales Badgerman)*. This tickled me, and made me think that there must be self-proclaimed Badgermen all over the country of Wales, each staking out their niche. (Was there ever a congress of the north-, south-, east-, west- and mid-Wales Badgermen?)

But the low-budget film was attractive. Here was a man who clearly loved badgers, who shared a very special bond with them and delighted in talking about them (little did I know quite how much). He might be the perfect guide and ambassador, I thought. I decided to give him a call.

I asked Gareth when would be best to see badgers and his answer was swift: 'Now,' he said. I did not have a lot of 'now' in store, but managed to scrounge a couple of days before the end of July.

The mid-Wales badgerman lives, unsurprisingly, in mid-Wales. But an indication of the strained relationship between badgers and man is that I cannot say exactly where it was in mid-Wales that we ended up. I began my journey from Oxford exhausted, and had to be fuelled by caffeine, and by the time I reached Gareth's home in Newtown, I was feeling awful – twitchy and sick.

Gareth was planning his seventieth-birthday treat and was bouncing around his house, making tea, spreading an inch-thick layer of butter onto a scone for me and talking, talking. He had a dozen things going on in his head at any one time, and the energy of someone half his age. His birthday treat was a new DVD of his 'family'. He was keen for them to appear relaxed, unaware of the camera. This was taking time, he explained: they just wanted to play. This meant he had had to step back a little, which was a shame. I had wanted to get really close to them.

His 'family', after all, were badgers. This might suggest that Gareth was lacking in human compassion; but he seemed pretty grounded. There were pictures on the wall of his neat little home that indicated he had a human family as well; and then in came his wife, Marian, and she normalised Gareth all the more. She was a straight-talking, no-nonsense sort of person. They had four children and seven grandchildren – and while Gareth talked about the badgers, she worried about their son who is serving with the RAF in Afghanistan. She had stuck with the badgerman for forty-one years, so far, sharing him for the vast majority of that time with his other family.

The layer of Welsh cloud had been gradually evaporating as we chatted, and the sun was beginning its slow descent: time to go and meet the family. A short drive, a short walk and then we stopped at a gate to absorb the scene. The valley before us was a bucolic idyll. The village church, the roistering rooks preparing to roost and the cows. 'Oh, the bloody cows,' muttered Gareth as we walked up the lane on the south side. It's not that they are any threat; it's just that they can be so curious that they interfere with the badger watching.

The richness of the evening light made the valley look like it had been manipulated in Photoshop – the colour saturation boosted to a point where it looked almost unrealistic. We stopped and watched; the cows drifted down the field and away from the badgers' sett, which was very conspicuous now that Gareth had pointed it out: between two magnificent oaks there was a patch of exposed earth. We were too far away to see if there was anyone out yet (having started humanising these creatures I couldn't stop).

We walked a loop to the top of the valley, allowing us to approach the sett up-wind to reduce the chances of my alien scent scaring them. Gareth's scent was very familiar to them, though sometimes they liked to top it up a little. 'You know, Hugh,' Gareth said, 'there are times when they do things that take me by surprise, even after all these years. Not long ago I was watching them playing when Scratcher came over to me, turned his back to me a little and then "pssst", right onto my trousers. He was marking me with his scent, marking me as one of the family.'

Scent is very important for badgers. They are part of the weasel family — also known as mustelids — which as well as the weasel includes the stoat, the polecat, the pine marten and the otter in the UK. All mustelids have musk glands, which they use for communication, and, in one particular case, for defence — for another species is the skunk.

We humans might not be able to learn much about the badgers in an area from their scent, but there are other ways of finding out what is going on. And just before we climbed over a gate into the field at the top of the valley Gareth stopped and pointed to the hedge bank on our right.

There was a path through the steep vegetation, worn deep into the ground over many years. Gareth pointed at it, then across the road; and there, sure enough, the path continued, down into the field we were about to enter. The path was small, only six inches across, but as we crouched down, we could clearly see some white hair stuck on a stem of bramble that bridged the furrow. 'Badger hair,' Gareth said, reaching forward and delicately picking one off. 'You see, this is probably under-fur: on the badger's back the guard hairs are mostly black.'

We climbed over the gate and into the field. Gareth took me straight to the path, which continued through the hedge. He crouched and gestured to me to look. To begin with I could not make out what I was supposed to be seeing, but then it became wonderfully obvious. There was a path of bent grass through the field. Just like the neat lines on my father's lawn, all the stems leaned one way, pointing in the direction taken by the last badger that passed through.

'That was last night,' Gareth said, confidently. 'But that one,' he added, pointing to another previously invisible line in the grass, 'that one is a few days old. You can tell, as some of the grass is springing back upright.'

I wondered how far those paths stretched. One of my hedgehogs walked over two kilometres in a single night, and having followed them, I knew that the actual distance covered by those little legs, as they followed their snouts from slug to snail to worm each night, must be far greater.

And it seems that badgers have a similar tendency. 'I always say that the badger can walk five or six kilometres in a night while never moving more than a few hundred metres from its sett,' Gareth explained.

There are more similarities than one might think between hedgehogs and badgers. Despite the obvious difference in size, they share a principle food source: earthworms. So for the most part, hedgehogs and badgers are competitors, the badger needing to eat seven or eight times as much as a hedge-hog. And here is where it gets interesting from an ecological perspective, and depressing from a personal one: when times are hard the relationship changes to one of predation.

And while it is lovely to think of a pack of hungry hedge-hogs bringing down a badger, obviously it is not quite like that.

Indeed the link between these two species is complicated. It is known as an 'asymmetric intra-guild predatory relation-ship'. And it may be that this relationship has contributed to the recent slump in hedgehog numbers. In research I have been involved with, we have found that there is an inverse correlation between the number of badgers in an area and the number of hedgehogs: the more badgers, the fewer hedge-hogs. This could be easily written off as a case of badgers simply eating hedgehogs. But they have happily coexisted for millennia, so that seems unlikely. What is more likely is that the food they are both after, macro-invertebrates in general and mainly earthworms, is becoming scarcer due to changes in the way the land is being managed.

The state of the land is something that got Gareth exer-cised. 'How many species do you see?' he asked. I looked. I feel a bit of a fraud as an ecologist; I know my hedgehog, but when it comes to grass, well, it just looks like grass. 'Just one species of grass, perennial ryegrass; and white clover,' said Gareth. 'I detest silage. I think it is the worst thing that happened in farming, because that's why we have no flowers. It's cut *before* the seed head comes on, you see. That is why we have no birds nesting in the fields, no curlews, lapwings, partridges; they are disappearing because the farmers are

cutting too early. When I was a young one you always heard
the curlews and the lapwings. Now there are no herbs, just
grass and clover. Grass is supposed to be the carpet of the
earth, but this is not a good carpet. This is like some awful
artificial fibre, it's like Astroturf. And then they come back in
with their little white pebbles that I call hail stones, and force
it again for another season – and what good is that doing to
the ground?'

And he has a very good point. The soil is becoming little
more than a structure in which to grow grass for silage,
induced by the application of those 'hail stones' – white
pellets of fertiliser. Soil is no longer allowed to be part of the
system for sustaining life. But why was this so upsetting to a
man whose obsession was badgers?

Badgers get TB and can give it to cattle. Cattle get TB
and can give it to badgers. There has been a vast amount of
work on the issue. It is not simple. Even down to the origins
of it all. *Mycobacterium bovis* is the bacteria responsible for
all this misery. And the word '*bovis*' shares a route with the
word '*bovids*', which is the term given to the family of cloven-
hoofed mammals that include cattle. Because it was cattle
that first gave TB to badgers. Now badgers act as a natural
reservoir for the bacillus, so that when it has been removed
from a herd (by killing all those affected), it can still return.
And how does the badger infect the cow? Even now, after
all the research that has been done, we are not sure. Could it
be urine? Or faeces? Or breath? And where does infection
occur? In the field? In stalls? In the feeding areas?

It's important to also look at this from the farmer's perspective. Many farmers are held in near-slavery by supermarket

chains that bind them into contracts in which milk is sold to the retailers at a price below the cost of production. Okay, there are some extremely wealthy dairy farmers out there, but they only exist at the expense of the smaller farmers, many of whom have been driven out of the business by the impossibilities of making a living.

So some farmers feel like the world is against them and, predictably, they scout around for something to blame. One of the additional expenses of dairy farming is bovine TB. Perhaps if this cost was no more, goes the logic, farmers might be able to break even? But how to get rid of TB? Well, since there is a correlation between the presence of badgers on some dairy farms and the presence of TB, the answer to some seems obvious: get rid of the badgers.

But in 1992 badgers received protection under the law. The lobbying by farmers has been intense, and the call has been going out for years for local, regional and national culls of badgers. At the British Hedgehog Preservation Society we have received letters from members of the National Farmers Union of Wales asking us to join in the call for a cull – after all, we know what badgers do to hedgehogs, don't we?

There have been extensive trials and massive reports – which have concluded that, while local extermination of badgers will help reduce bovine TB, this will not be a permanent fix, and has the added risk of driving TB-infected badgers further afield.

Back to the grass (and this is related), Gareth is not alone in believing that the root of the problem lies not with the badgers but with the way that livestock and the land on which it relies is managed. 'Dairy farming is just too intensive,' he explained. 'There are too many cattle on the fields, the fields never get flowers and herbs growing, the mineral content of the soil is diminished, there are fewer insects and birds, the cattle have to spend months at a time on concrete inside during the winter, they are forced to produce more milk than is good for them and are given antibiotics just to keep them going. And the farmers wonder why they have poorly cattle. It makes me so bloody mad. They are forcing their animals to do things that they shouldn't and then blaming the poor badger when it goes wrong. Well, the whole lot should be broken up [make the farms smaller and less industrial and start paying farmers properly for the milk they produce]. We need to sort out this bloody problem and not just go killing badgers because we are too scared to look at the real issues of corporate control of the food chain.'

I was taken aback by his vehemence. The reason that Gareth had his dander up was that he had been involved in the campaign to get a proposed badger cull in north Pembrokeshire stopped. When we met, the news had just broken that the Court of Appeal had declared the cull unlawful. The frustrating thing about plans to cull badgers is that it misses the point. It might have a short-term effect for the dairy farmers, but the long-term effect of continuing to farm in this manner will have far worse consequences.

Graham Harvey may be best known as the agricultural story editor of *The Archers* on BBC Radio 4, but he is also an eloquent writer, who has tackled some of the most fundamental problems facing agriculture. For example, in his book *We Want Real Food*, he argues that we are 'fifty years into a mass experiment in human nutrition. We're all eating basic foods that have been stripped of the antioxidants, trace elements and essential fatty acids that once promoted good health.' Those 'hail stones' come in for similar criticism. Fertilisers, he says, are the 'powerhouse that drives modern farming'; they have 'degraded our everyday foods and led to the upsurge in ill-health'.

As for people, so for badgers. For now, though, Gareth's family is safe, and the owner of the land where the badgers live, who has tolerated Gareth for all these years, is obviously not too concerned. When we finally got to the sett I could see evidence of cows all around. They coexist just fine here.

I was worried that all our talking might have scared the badgers off, and Gareth did admit that we probably wouldn't see the more timid animals, certainly not this early in the evening. It was nearly 8 p.m. He asked me to stay where I was, and to keep quiet. Getting up, he pulled a large bag of peanuts from his rucksack.

When he first came to this sett, it was not to see badgers. He had seen a nightjar flying nearby and wanted to get a closer look. And as he settled down to look for the bird, up popped a badger's head, and another; little teddy-bear ears atop the easily recognisable black-and-white-striped face. Gareth was hooked. He went back the next night with some Chelsea buns, trying to inveigle his way into their hearts (or stomachs). They did not go for the buns, but when he tried peanuts, well, it all changed (though he did admit that their absolute favourites were digestive biscuits and Mars bars).

So while I sat beside a patch of nettles, on earth scraped bare by countless badgers, Gareth sprinkled handfuls of peanuts on the ground with a sound like rainfall. And then it seemed as if Gareth disappeared into his own world. Like a shaman chanting, he started talking to the badgers underground. 'One, two, three, yahhhh, here you are, one, two, three, yahhhh, come on, it's only me. One, two, three . . .' And on he went, oblivious to the rest of the world.

As he settled back down with me he put a finger to his lips and pointed to his ear. We both listened intently, and then it dawned on me what the sound was – almost as soon as he sat down it had started. Lips smacking, teeth cracking peanuts, as someone started work on those that had fallen into the entrance holes.

Then I saw some movement: grass near the sett entrance moving. And then a snout . . . a pause . . . another snout. As the first badger emerged to start on the peanuts around the mouth of the sett, Gareth leaned over to me and whispered, 'Now, Hugh, that first one is a boar, three years old, and I don't have a name for him. I'm whispering because that next one is a cub and it might take fright. But it's not Tiny's cub – and there *is* Tiny; she's been with me for nine years now – you can see she's going grey like me. But she's in very good condition for her age.'

The three animals were about ten feet away and utterly uninterested in us. They seemed much smaller than I remembered from previous sightings. Gareth explained, 'When it's dark, then something strange happens. They always seem bigger in the dark.'

There have been reports of badgers up to the size of small bears. Which, given a healthy dose of alcohol, a vivid imagination and a prone position on the ground looking up at the approaching monster, is easy enough to imagine.

In fact the usual weight for badgers in the UK is around 10 kg, which is about what you would expect a beagle to weigh. But somehow the badger seems heavier. Gareth has had some big ones at the sett over the years. 'Scratcher, he was a big badger; and Barnaby, he was my enemy, he was jealous of me, he always reminded me of a Labrador on very short legs, he was a massive thing.'

Gareth may have had an enemy in Barnaby, but his relationship with these animals has, on the whole, been very positive. Yet of all wildlife in the UK, the badger has to be the species least likely to make it into the cuddle-off for cutest critter of the year. How many animals have become synonymous with such negative attributes? 'To bait, hound; to subject to persistent harassment or persecution.' That is the view of the lexicographers of the *Oxford English Dictionary*.

Then again, while the badger may come with plenty of baggage, there is no denying how well rooted it is in popular consciousness. The badger is a regular feature of children's books, from *Wind in the Willows* to *Percy the Park Keeper*. There are badger groups all over the country, and the Wildlife Trusts organisation has the badger's face as its logo. So they are well attended with public-relations support.

There remains, therefore, a contradiction in our relationship with the badger that is more profound than with any other British animal. Badger groups and badger-watching holidays, story books and cuddly toys are set against a very different narrative – one that, at its most extreme, has resulted in the deaths and torture of thousands of these animals.

The idea that there is 'sport' in killing animals for entertainment is not one with which I can sympathise. It is a fairly simple ethical question and I am always surprised when intelligent people get caught up trying to defend cruelty for fun. Both bear baiting and badger baiting were banned in 1835, but badger baiting continues. Even in the twenty-first century there are people who get their kicks from seeing dogs and badgers rip each other to pieces.

The first book about badgers I read was *The Darkness is Light Enough*, by the pseudonymous Chris Ferris. Gareth had met the author and was hugely impressed by the strength and courage of this diminutive woman. She would walk the night through, keeping an intimate diary of her encounters both with the wildlife and with the vicious thugs – badger baiters – who hospitalised her as she dared confront their cruelty. That was in the mid-1980s. In 1992 the Protection of Badgers Act was introduced, yet still badger baiting continues. The law demands severe punishment, but enforcement is woeful, with the vast majority of prosecutions being obtained not through police investigations but through the dedicated and brave undercover work of badger fans.

Even in this mid-Wales paradise, Gareth was cautious. He did not advertise the presence of the badgers but he had mapped all the setts in Montgomeryshire and shared that information with the police. The extreme cruelty of the baiters is not the only threat these sturdy animals face from people. Despite the legislation protecting badgers and their setts, there are still farmers who would rather not share their land with Brock.

Gareth had seen it all. 'I don't want to run landowners down, there are some excellent landowners. But there are some who are bad. They shoot badgers, they put poison in potatoes or tennis balls and throw them into setts, or they pour slurry down the setts.'

And then there were the road casualties. 'But at least they are accidental,' I said. But Gareth was deeply suspicious of many road kills. 'I started to get worried when I found badgers that had apparently been run over, full of shotgun pellets. So now I pay more attention. If you find a badger on a road, have a look. You remember that badger path I showed you coming down onto the lane back there? Well, if there is no sign of a well-worn badger-path leading to or away from the dead badger, be suspicious. If there is no bleeding nose there is something wrong as well. Farmers kill badgers and then dump the bodies on the roads to make it look like an accident.

'You look at these now,' he went on, 'just a few feet away, happy, eating peanuts. These are intelligent and caring animals,' Gareth said. 'They are not just obstacles to be killed and discarded. They treat each other with more respect than we treat each other.'

Sometimes people will talk about being madly in love with an animal, but you can tell that it is a superficial and somewhat sentimental affection. But when Gareth talks about his love of his badgers, it is indistinguishable from the way he talks about his wife, children or grandchildren.

'Emily was fifteen when she died. I named her after my youngest daughter; I brought her up here to meet the little sow and named her then. She was so very close to me; she used to guard me. She would always stand beside me and if she were here now and you moved she would growl – *grrrrhhhhhhh*!' Gareth has slipped into reverie. 'And then there was Scratcher; oh, I could do anything to Scratcher. People still write to me and ask if he is here. He loved human beings. I taught him to love humans, and I cried a cupful when he died. He died so suddenly, he was fifteen, but did not look old. When I came to the sett he was always the first one to me and this time he did not come. He was standing up there and I called and he looked up at me and I said, "What's wrong, you not hungry?" and he looked around at me and walked down the bank – oh, I can see him now. That was the last time I saw him alive. I found him the next evening and I had lost a friend.'

By now Gareth is crying, quiet tears working their way through the creases of his weathered face. 'Emily was the same; I miss her too. I was here shouting, "Em, where are you?" She always wanted me to feed her on her own, not with the others. And I wanted to see her for another spring. She came out in the spring and she came straight to me and I said, "You alright, Em?' She was blind, she had cataracts, just like an old dog, and I went home and told Maz that Em was there. But I never saw her again. Not alive anyway. I was sitting here a few weeks after and I could smell dead badger; I looked down, just where your foot is now, and there she was, all curled up like a cat and I left her there. She was left there until she went to bones. No badger went near there.'

We both have tears in our eyes now. His love is infectious. 'I think that badgers have sort of healing powers,' he said, surprising me. Sensing my hesitation, he went on, 'Okay, there was a woman who had been in a wheelchair for four years; she walked back to the car after seeing the badgers. And a man was dying of cancer; well, I brought him here as a last wish. He calmly opened a pack of biscuits and placed them on his lap, and the badgers were up and on his knee.

They were leaning on him as he wept. When an animal can do that for a person and we treat them so poorly, that is so wrong.

'I love them like I love my wife,' he said, without a hint of hyperbole. 'I am not saying "more", because that would be dreadful, really. She is pretty good, Maz, my wife. Sometimes she will ask me not to go up on a night, but it is like a magnet, something pulls me, and I say, "We've brought up our family, Maz, and I've got another family up there to look after." She copes with me, but she thinks I'm *setts mad*.'

It took me a moment to realise what he had said; his eyes were twinkling with mischief.

'They all know their names, you know,' he continued. 'Miranda, she was semi-albino and she would come and put her paws up on my knee. She disappeared; she was old. And then one night there was a terrible storm and a flood of water cascaded down the hill and blew a skull out of the hole, just there. It was Miranda's. She always had one fang missing. Then there were Upsups, Sput, Young'n, Barnaby, Seenie. You know, I once caught Emily and Scratcher copulating here!'

Now Gareth was grinning again, tears dried. 'That was great fun; you never see that much of it, and they were here having a go at it. Well I didn't know what was happening to start with; Scratcher was there making this noise – *woowoo-woowoo* – and I said, "What's up, Scratcher?" He paused and looked at me and then suddenly up popped Emily's head, all covered in muck and he was back at the *woowoowoowoo*. They were at it for ages, I even tried offering him some food, but he was having none of it.'

In the space of a conversation Gareth has gone from rage to grief to glee. But underlying it all is love. This is a tricky subject. I often find myself talking to a group of people, whether it is the WI or a post-graduate research group, and will get to the point where I have no other option but to use the 'L'-word. There is no other word that conveys the deep and complicated relationship it is possible to forge with wildlife.

But there should be a word. My friend Roman Krznaric, author of *The Wonder Box: Curious Histories of How to Live*, wrote that the Ancient Greeks identified six different varieties of love: *eros* (erotic love), *philia* (the love of friends), *ludus* (playful love), *pragma* (pragmatic love), *philauteo* (self-love) and *agape* (universal love). It is absurd that we simplify this rich diversity into a monoculture. Like the fields of perennial ryegrass, it does not do us any good at all.

'We've got to fall in love with nature,' Gareth said as we walked back towards the car. The moon was just off full and loomed large above the hillside. It was hard to imagine not being in love with this tranquil corner. 'And my badgers and your hedgehogs, they are like gatekeepers to the wider wonder of the natural world. I bring people down here at eight o'clock for an hour and I find I am still here at one in the morning. I get the barn owls quartering the field, probably hunting for the woodmouse that sometimes sits on my knee. And then there was the stag beetle that would come and take peanuts, one at a time. You know, there was a blackbird who would sit on my shoulder and a chaffinch who would follow me from the car to the sett where he would wait for a peanut.'

As we got to the gate we paused and looked back down

into the darkening valley. 'We should be in love with nature; it's all we have got. I'm coming on seventy now, and I'm not going to be here soon. But for my children and for theirs, we have to do something.'

It was not clear if he was talking about his children or his badgers now, but it doesn't matter. We stopped and sensed the wonderful world around us, before heading back to Gareth's other family – and another scone with a wedge of butter on top.

3. BATS

'A blur of shadows, pips and squeaks'

'If someone were to bomb the National Bat Conference,' I was warned, before embarking on this chapter, 'Britain would be in danger of losing its reputation as a hotbed of eccentricity.'

Huma Pearce was fierce, green-eyed and passionate. When we met near a park in west London and I told her about my idea for this book, this passion was manifested by her determination to win. She saw it very much as a challenge to bewitch me into the world of bats.

Huma was an easy choice of advocate. I remembered sitting next to her at a Mammal Society conference and noticing a bracelet of bats tattooed around her right wrist. That took some confidence, especially in the relatively conservative world of science. I knew, because of the looks of amazement I got when I rolled up my trouser leg to reveal my hedgehog tattoo.

Huma's combative spirit has enabled her to establish her own ecological consultancy, Mostly Bats, and to forge a career in an area she loves. It also meant that she went an extra mile or two in preparing to try and captivate me with her love for bats. She was taking no chances.

Dusk was creeping in around the margins of Chiswick House and Gardens. The park was emptying of dog walkers as we entered. More than one person warned us that we would get locked in if we did not turn back. But that was the idea.

'I like being alone,' Huma said as we made our way deeper in, and the drone of the M25, the M4 and the A4 began to be muffled by the trees. I felt, almost, like an intruder – not an intruder in the park so much as an intruder into Huma's world.

Our target was a potential bat roost. But it was late August and Huma was muttering about my timing. 'Two months earlier and I would be able to guarantee bats at this roost, but now, well, this is the end of the season.' At last we stopped, and Huma began to pull out her bat-detecting equipment. 'This is a heterodyne and frequency-division detector,' she said, as she handed me a box of electronics about the size of an old Walkman. 'Put the headphones on – but don't plug them into the headphone socket. Pop them into the other hole and set the frequency to forty-five kilohertz.'

The bat detector is a wonderful piece of kit. Bats are identifiable by the high-frequency noise they make as they fly and hunt. This is known as echolocation, and it allows them to get a picture of their world without the need for light. Most importantly, it enables them to pursue prey, even through dense foliage. But the noise is too high for us to hear – the frequency we humans can detect peaks at around 20 kHz in children. (KHz, or kilohertz, is the measure of frequency: the higher the frequency, the higher the note.) For example, the note an oboe plays at the beginning of a concert, for the orchestra to tune to, is an 'A', at 440 Hz or 0.44 kHz.

These bats were squeaking at 45 kHz, so we wouldn't hear them without a little help from this box of tricks, which generated a representation of the calls at a frequency even my aging ears had a chance of hearing.

But at the moment? White noise. The sound of no bats, I suggested. Huma flashed her eyes at me and turned back to her machine. And then I jumped as I heard the first intrusion into the white noise – a high-pitched, rapid pulse. 'A bat,' I declared. 'My zip,' Huma replied.

Her headphones were big, DJ-style, whereas I had a small bud in each ear. The technique, I learned, was to keep the right earpiece 'out', and listen with that ear to the speaker on the detector. 'With the detector set at forty-five kilohertz, if a pipistrelle bat were to fly by, you would hear its calls through the speaker, but if a noctule bat, which echolocates at around twenty kilohertz, was passing by, you would hear this in your left ear piece.'

Dusk had descended into dark. 'Should I be quiet?' I asked. 'No, they're not coming out now. Or they did already, before we arrived, or they were not here at all. I'm going to have a fag.' And there was another strange noise in my ear as she struck a match. Which was nothing compared to the high-pitched screech that made us both flinch when I touched my camera as I was putting it away and the auto-focus kicked in. Not quite the silent machine I had thought it was and a reminder that there are sonic worlds out there all the time of which we remain entirely ignorant.

As we walked away from this bat-free zone, Huma explained a little more about what we were hoping to hear. 'Bats have two types of calls. There are constant-frequency calls and then there are frequency-modulated calls. Constant is just like hitting the same note on the piano repeatedly; the note gives an indication of the species. But the modulated calls go from high to low frequency, so on a sonogram there is a *vertical* line for these bats, whereas the line is horizontal for the constant-frequency calls. You got that?'

A sonogram is a graphical representation of sound.

'Of course, it is never that simple – and most bats use a combination of the two. For example, a pipistrelle or noctule call will show on the sonogram as something like a hockey-stick shape. Their frequency tends to be more constant in the open and more modulated amid foliage, for instance, to give them a more detailed picture of what is going on around them.'

The detail that these echoes can pick up is extraordinary. We are so used to a world of sight, where light is reflected back off our surroundings and translated in our brains. This restricts us. When there is no light we need to use a torch. And bats? Well, they can see as well as you or I with their eyes in daylight, but when it gets dark they have the equivalent of a torch built into their heads. In the dark they don't use light to capture reflections; they use sound.

'Think what these amazing animals can do, right,' said Huma, 'and then I keep finding people getting all agitated about them getting stuck in their hair. For goodness' sake!' Huma is not always very tolerant of the ignorance of others. 'If they can snatch unsuspecting insects from leaves or grab them mid-flight, they're not going to get tangled in your hair.'

As we walked on towards a small river, I suggested that London was not exactly the most bat-friendly of places.

Perhaps Huma might have done better taking me to somewhere a little more, well, impressive.

'There are bats all over the place. I wanted to show you how amazing they are, fitting into our hostile world. I was sat in a pub in Brixton once, having a pint, and my bat detector picked up pipistrelles.

'There were other people there as well,' she said, somehow sensing my raised eyebrow. 'Pips are house bats. They used to live in trees, but we came along and chopped all the woodland down and they found that our shelters were a pretty good replacement. So, old farmhouses are great, where they can squeeze themselves in behind the mortice joints. Even modern buildings can be fit for bats. You could have a hundred pipistrelles roosting behind a facia board, and I counted over 700 leaving a cavity wall in the outskirts of London.'

Pipistrelles have learned to adapt to our environment more than any other species of bat. I wondered whether our artificial lights might be of assistance to them. After all, if you want to attract the bat's food, moths and flies, all you needed to do was leave a light on.

'After habitat loss I would say that the major obstacle bats face is light. Some species, like the pipistrelle, can fly into the open and are not massively impacted by light, even though it will increase their vulnerability to predation. But most will not, and the insects that the street lights attract are being drawn away from their natural environment, further affecting the ability of other bats to thrive.'

As we approached a bridge I was about to ask what had prompted her interest in bats, when Huma said, 'There was a bat, did you hear it?' Not a good start – I was too busy talking. All I could hear now were police sirens. We reached the bridge and almost immediately it sounded like I was being strafed by a machine gun. A rapid fire of clicks, shooting from left to right. My first detected bat. And suddenly I was much more excited. I could see shadows flitting as the receiver in my hands sputtered out a succession of unearthly noises.

'They're feeding over the water. Pips – forty-five and fifty-five kilohertz,' she explained. For a while we just enjoyed the spectacle. In the city there is enough ambient light to see quite clearly, and the white stone of the bridge provided a perfect backdrop to the erratic dance of the hungry bats.

Like farmers on a gate, we leaned on the bridge wall, gazing over the water and chatting. The bats were undisturbed by us, fixated as they were on the wealth of midges flitting up and down over the water. I asked her again about her interest in bats.

'My passion has been small mammals for as long as I can remember,' she said. 'As a child my bedroom was shared with cages of rats, gerbils and hamsters, and while I was at university I travelled to Borneo to research weird and wonderful rats and shrews. Things got a little bigger during my Master's degree project, when I studied otters. I worked as an Otter Officer in the UK for a couple of years, but got itchy feet again and upped roots to Peru in search of giant otters. The one constant, though, was that wherever I was, I would always see bats.'

Unfortunately, money ran out and she had to return to the UK to find a 'proper' job.

'Five years later I found myself in a desk job, ploughing

through planning applications with respect to protected species, and that's when the bats reappeared. So many of the development proposals I reviewed had adverse impacts on bats, destroying their roosts or covering their feeding areas in concrete. I wanted to do more to help, so I phoned this guy, a bat trainer – he trained people, not bats. You have to have a bat licence to disturb or handle them, and you need to be trained in order to get one. He asked what I was doing that night and I said I was going out bat hunting with him . . . And that was it. The first time I held one of those little shadows in my hand I thought my heart would pop. So beautiful, so mysterious, so vulnerable, and there it was in my hand.'

Since that moment there had been no turning back. Huma's love of bats had manifested itself in a bracelet of bats tattooed around her wrist and a life dedicated to helping these animals.

A large part of that involves education. There are still many myths to be scotched about bats, and not just the one about them getting tangled in hair. 'Some people think of them as blood-suckers that will come for them at night, others think they build nests and will damage their property by roosting in their attic. The most frequent misconception is that bats are flying mice. Bat roosts are frequently misidentified as mouse infestations, based on their pooh. A builder will be in someone's attic and emerge saying there's a massive mouse problem. But all you need to do is pick one of the droppings up and you can easily tell the difference. Mouse droppings are smelly and sticky; bat droppings are dry, largely odourless and crumble between your fingers.'

Time to move on. Huma had bigger bats in mind. 'There are several different sorts of bat calls, you know.' This is what I love about people like Huma, what passes for a normal conversation is actually a gentle lecture. 'As well as the typical calls they produce during their nocturnal flights, you'll also hear a variety of social calls and feeding buzzes. The feeding buzz is – oh, you're recording this, aren't you? Well, the way I describe it is – I'm embarrassed now – okay, you know how it is if you are trying not to fart and you clench and a sort of squeak comes out?' By now I was nodding in the half-light, hoping that Huma hadn't noticed my grin. 'Well, that's like the rapid pulse of sound waves the bat gives off as it comes in for the kill. More pulses is a bit like more light – it gives them more detailed information about what is there. It's known as a raspberry. The social call is a more relaxed, low-frequency sound.

'People think of bats as flying mice,' Huma went on, once she had regained her composure. 'But they are more like *us* than like mice.' That will take some explaining, I thought. Sensing my scepticism, she added, 'In lifestyle, at least. They live for a long time, up to thirty-five years, and usually only have one pup a year or every other year. They invest a lot of time in nurturing and teaching their offspring about the best roost sites and feeding grounds. They are very social animals, and this information is passed on from generation to generation. A roost site may be used by a colony for more than fifty years.'

And while mice are rodents, all UK bats are chiroptera, and are all highly insectivorous. And they certainly are pretty voracious when it comes to those insects; they can consume up to 3,000 midges in a single night. Feeding at that rate, or anything near it, means intensive work. Catching four or five insects per minute does not leave a lot of time for anything else. This feeding-rate also reveals the fragile balancing act of bats' lives.

Flying enables bats to fill an ecological niche left largely vacant at dusk by the usually diurnal birds. But it requires a lot of energy, so they need to feed continually while they are active. They have to match the amount they are able to eat to the energy they expend. And over millions of years they have got the balance just about right. Changes happen, of course. Nothing stays the same. But, in nature, changes normally happen slowly enough to allow species to adapt.

But slowly is not how we humans alter the environment. An indication of how subtle this balance is comes from work that Huma is doing with Alison Fure of the London Bat Group. They have been looking at Daubenton's bat, a species known to be sensitive to light, and seeing how it behaves firstly when there is no moon and then when there is a full moon. The full moon, on a clear night, generates about one quarter of a lux (the standard measure of illuminence). And this was shown to reduce Daubenton activity by 63 per cent. By way of comparison, a very dark, overcast day produces about 100 lux, office lighting 500 lux and direct sunlight over 100,000 lux.

The point is that if the bats behave differently with an increase in light intensity of just a fraction of a lux during the full moon, then they are going to be seriously affected by a spread of manmade lighting. Already researchers have shown that the relatively insensitive pipistrelle will alter the routes it takes when lighting reaches 3 lux. So a pub with a new outside light, for instance, can fragment the bat's habitat, restricting its ability to move around.

While cutting down artificial light might suit the bats, surely a balance had to be struck. For instance, artificial light makes paths safer. But Huma was ready for that argument, 'You can use more sensitive lighting,' she said. 'It's quite straightforward, really, and it saves money, too. So much of the light in use at the moment does nothing more than light up the sky. It would help if lights were lower, and aimed at the ground they're supposed to illuminate. Better still, in really sensitive areas, like alongside canals and reservoirs, use motion-sensitive lights. You don't need to keep a path lit all night long for just two dog walkers, do you?'

Sometimes Huma was hard to argue with. And the disturbance caused by artificial light is not the only concern Huma has about her urban bats. The insects on which bats feed need somewhere to live, and that means vegetation. London is not over-burdened with gardens, despite having a great selection of parks. One solution to this lack of vegetation, favoured by many ecological consultants, is to install green, or 'living', roofs – actually, they are not called that any more, as Dusty Gedge, living-roof guru explained to me once. 'If you want a green roof, get some green paint.'

This is an area that Huma has been researching. The theory goes that if these living roofs attract more insects into an environment, then that should be good for the bats. But are living roofs really better? Some may say that it offers a way for developers to 'greenwash' an otherwise insensitive building project with a sprinkling of sedum toppings.

But how to find out what is going on? Was Huma sitting on roofs all night long, clutching her detector. 'Strangely, people are reluctant to leave me on a roof overnight,' she retorts. 'I have automated machines I leave on roofs over the course of a week, and tomorrow morning you are going to help me collect them . . . A noctule! Set the detector to twenty kilohertz!' came the sudden instruction. And then, there up between the trees, something much larger than the pipistrelles was getting Huma excited. To start with, all I could see was a plane and its condensation trail. But a few moments later there was something I would have dismissed as a bird, were it not for the little imp at my side almost skipping on the spot.

I switched to the lower frequency. There was something very different going on in the air – chip, chop, chip, chop. Like a ticking clock. Much more sedate and single-minded than the pip and its figure-of-eight antics, the noctule was homing in far more directly on larger insects like cockchafers. Then a second noctule joined the fun and as I followed them I almost toppled over, so focused was I on their flight.

As I gazed upwards, I felt the first drop of rain. Heading back to the gate, Huma revealed how determined she was that I end up with a bat tat. She had arranged an adventure for tomorrow afternoon. We were going to meet a very special person, who could 'win over anyone to the side of the bat'.

Well, that would be quite a challenge. I have grown fond of hedgehogs partly on account of the opportunity they presented for close contact – looking into the eyes of a wild animal feels like a key means of connection. And while the bats had put on a spectacle and were clearly important species that even the most urban of dwellers could have some link with, they remained little more than a blur of shadows, pips and squeaks.

'They are addictive, you know,' Huma interjected into my

thoughts. 'Once you start, it can be hard to switch off from bats. I always warn people, it ruins your sleep and your social life, and it ages you fast – it's as good, or bad, as a drug. The trick is to hibernate to recuperate.'

And then there is the iconography of the bat. A more immediately captivating animal image is hard to imagine. DC Comics have done great service to making the bat 'cool', perhaps undoing some of the damage that Bram Stoker did with Dracula. The bats tattooed around Huma's wrist are 'Batman' bats.

As I stretched out on Huma's sofa for the night, a magnificent bat-ish painting leaning against the wall beside me, my mind drifted into a playground debate: Batman vs Dracula – who would win? I was hoping for some insight as I slept, but none came.

Next morning, after a super-strong cup of coffee, we were off again. This time to see what rooftop messages the bats had left Huma in the last week. On the way I mentioned the Batman/Dracula question. I don't think Huma was very amused, but she did point out that in terms of doing good for society, in reality vampires probably win hands down. Vampire bats, of which there are three species in Central and South America (not Transylvania), do feed on blood but not by sucking. They cut the flesh of, usually, cattle, horses, pigs and birds, with razor-sharp teeth and then lap at the flowing blood. They are small animals, up to 9 cm long, and take about a tablespoon of blood a night, so are unlikely to drain a sleeping human.

And while there is the risk of rabies, there is something rather beneficial about this bat. Their saliva contains a chemical that stops the blood clotting – an anti-coagulant. And it is a remarkably effective anti-coagulant that is now prescribed for heart attack and stroke patients. They had to give it a name … Draculin.

Batman, she reminded me, is fictional.

We collected the bat detectors from the roofs of two properties – one covered by grass and wild flowers, the other with more conventional slates. Back on the ground, Huma spent some time downloading data, checking batteries and plotting our next adventure. The detectors were on the move, as they will continue to be until she is satisfied with the spread of data. She showed me the sonograms for the pipistrelles and noctules that had been recorded. She was delighted, but I still felt I was being offered shadows and echoes. And while I was fascinated by these suggestions of life, they did not grip me like an encounter with a hedgehog. When I said this to Huma, however, she exuded the contented air of a poker player aware of the straight flush hidden in her hand.

Back in my car, she input a postcode into my satnav and looked at me expectantly. Knowing better than to ask, I started up the engine and we wove our way south into the West Sussex countryside.

Still so close to the Great Wen and almost suddenly, large trees erupt. There is a belt of wealth at just the right distance from the city that has managed to keep its old growth intact despite its proximity to London. And it was the heart of this band that we were entering. Down a pitted lane we bounced and emerged at a pleasant house, which had had its wildness protected by the nearby presence of a golf course.

By now Huma was almost beside herself. If this had been

a game of poker I would have long since discarded my hand. She had been plotting and was about to reveal all.

A wooden plaque on the gate declared the house to be the Sussex Bat Hospital, and we were greeted by Jenny Clark, who immediately gives an impression of strength in a delicate form. Not a million miles away from a bat, in fact. She was buzzing with an intense energy to match Huma's excitement, and I was rapidly feeling like a gooseberry as the two women started sharing bat stories. Huma came here once as part of her training, and my desire to be seduced provided her with the perfect excuse to revisit. 'There is no one who knows as much about bats as Jenny,' Huma declared with a great grin. And I just leaned back against the fridge and listened. Because I could listen to Jenny for hours – in fact, that's precisely what I did. She spoke with the precision of an earlier age, every word perfectly enunciated.

'The biggest problem is cats. And I am forever telling people that their dear Mr Cat has to be kept inside after dark. Mr Cat can hear the echolocation of a bat and is mesmerised by the sound.' Jenny has a mesmeric tone herself. 'This is not because they eat them. They don't, because they taste horrible. No, it is worse. They just play with them. You have seen how a cat will leap at a toy; well, they do the same with the bats, and snag on their wings with their claws. As far as Mr Cat is concerned the game is then over, but the poor bat is left on the ground in the night waiting for Mr Fox to come and finish him off.'

The bats most vulnerable to cat-attack are new mothers. 'When the baby is a week old it becomes too heavy for Mum to fly with, so she scrapes it off the nipple and puts it in the crèche with the other youngsters, and nips off to the super-market for a feed. Not once, but up to seven times during the night. Now the sad thing is, when Mr Cat, sitting by the pond waiting for his fun, tosses mum, who crawls away to die, well, she doesn't return home. And the youngster says, "Where's my mum? Because all the other mums have come back, but mine hasn't, so I had better go looking for her." So it climbs down the side of the house like Spiderman, running across the patio with its mouth open going, "Squeak, squeak, I want my mummy", to be found by the house owner in the morn-ing. And if they don't climb down the wall they go through the internal workings of the house and come out in the airing cupboard. You might have noticed that all the pipes lead-ing to the airing cupboard have a wee half-inch gap between them. And it is warm. And you open the cupboard to take a towel and you are confronted by a little pink bat, going, "I want my mummy". And the house owner says, "What shall I do?" and they phone RSPCA, and they say, "There is this mad lady who lives in Forest Row near East Grinstead, she hand-rears orphans, try her."'

Jenny Clark has to be one of the un-maddest mad ladies I have ever met. Eccentric? Yes. Obsessive? Definitely. Mad? Well, don't let the cutesy language fool you. All this talk about Mr Cat has a very serious purpose. She has long been campaigning to get cat owners to take more responsibility for the behaviour of their animals. Cat owners have a remarkable ability to disregard the carnage their pets cause to the ecosys-tem, shrugging it off or leaping into denial: 'Well, my little Tiger would never do anything like that'. Well little Tigers all over the UK are killing around 55 million wild birds every single year. And that is not a random statistic generated by knee-jerk.com, but the result of serious work from some of the best ecologists in the UK, at the British Trust for Ornithology.

Unlike me, Jenny has learned the technique of subtlety when it comes to dealing with this sort of issue, and has found a way to appeal to the self-interest of cat lovers. Vets have told her that most of the cats that are brought in have been in either a collision with a car or a fight with another cat, both of which nearly always happen at night. So the answer is simple. If you want to protect your cat, you need to keep it inside from an hour before dark to an hour after sunrise. Oh, and as a side-effect, you will also help save the lives of thousands of bats.

There are some people, however, who might be beyond reach. Jenny told us of a lady who phoned up with a serious bat/cat problem. 'Every night,' she said, 'my cat goes up the stairs and into my bedroom. I have a cream carpet. He goes to the open window and sits on the chair and at dusk, as the bats fly out just above the open window, he puts out a paw and hooks one in to play with it – and I don't like that happening because it leaves blood on the carpet . . . what are you going to do about it?'

The key to Jenny's work is education. When Huma first met her it was to gain experience with bats in order to qualify for the handling part of her bat licence. Jenny has an 'education team'. From May to September they are out at the Women's Institutes, primary schools, agricultural shows, zoological conferences and every other conceivable forum that enables them to promote the wonder of the bat.

'I do over seventy talks a year, but nothing from the end of September. I need to rest my voice and my bats, and catch up on all the reading.'

Jenny does not have much time for reading during the active season – or much else, in fact. 'My husband is very tolerant and supportive. I don't cook in the summer, everything is from the microwave. Look around; where would I prepare food anyway, even if I had the time?' And she was right. We were standing in her kitchen and most of the surfaces were covered with cages containing bats. 'This is A&E. When they first come in I have to make the big decision. All the time I have to consider, what is best for the bat? There are three categories. About a third of them I can fix and release. A few others never quite pass the flying test, and join the education team. And the remainder, well, they have to be euthanised.'

I was intrigued by the idea of a flying test and asked how it worked. 'They get to show their skills in the sitting room. I spread a duvet on the floor, shut the curtains and turn off the lights. Then they have to show me whether they have mastered three things: skills, stamina and attitude.'

I assumed she'd said 'altitude', but no, apparently attitude was vital. 'They might have the skills to fly properly and the stamina to stay up, but if they lack the right attitude, then back into the hospital they go. They have to be shouting "How do I get out of here? Has she left the window open? Is the door closed?" They always go around the door and window, examining the corners trying to find a way out. If it's "Flap, flap, where can I hide?" – wrong attitude. Lacking confidence, loss of nerve. So rest them, try them again in a week – try several times for several months and if they are not able to be released, they join the education team.'

And now we were going to meet the education team. Huma's excitement had ratcheted up another notch. It wasn't just the closeness of the bats that was getting to her; she was convinced that this would win me over to the world of bats. Then Jenny, apparently unaware of Huma's agenda, stepped in and guided us to a back room. 'Now, this is where you get to put the face to the shadow,' she explained.

My suspicions were rising. That choice of language: that was precisely how I had put it to Huma last night. Had they been conspiring?

'I am going to start with one of the big ones. Now, I am sure you are aware that bats are divided up into two sorts: hawkers, which chase insects; and gleaners, which snatch their prey from leaves. And we are going to meet a serotine. Now, where is she?' Jenny had unzipped one of her large cages and was rummaging up the leg of a pair of trousers that were hanging on the back wall. 'I ask all my friends to pass on their old trousers, especially corduroy. They make a good approximation to a tree. Got her.'

Firmly holding this bat in one hand, Jenny zipped back up the cage and turned and said, 'There are three large bats in the UK: serotine, noctule and greater horseshoe. Every species has a totally different temperament and personality.' And with a little flourish she opened her hand and there was the most adorable furry bundle. Thick, shaggy brown fur. My first instinct was to stroke it. 'Now, we mustn't touch,' Jenny said, seeing my hand approaching. 'Our skin has oil and bacteria on it and the bat must keep its fur pristine. It needs to be fluffy in order to trap air for insulation. I'm wearing cotton gloves and I change my gloves for each bat.'

Well, that told me, but I still wanted to stroke the bat. I had a soft toy hedgehog as a child. I think it was probably made of real fur because it was unbelievably soft and I would stroke it until I fell asleep while I was away at boarding school. I still have a bit of a fur fetish and probably need to get a very compliant dog.

'The serotine is a partial gleaner. It will take some insects on the wing, but most of those it catches it will grab from foliage. They are also fond of settling down next to a cowpat and feasting on dung beetles. The heavy doses of insecticides cattle have been fed to control internal parasites will not have been doing this species of bat any favours, because the chemicals make the cowpats toxic to bugs.'

As I was being told all this, Jenny was indulging in what looks suspiciously like the sort of fur-therapy I would enjoy. 'Bats love to be cuddled and stroked, and when you do it properly they will actually purr. Listen.' And there was the strangest noise of contentment emerging from this furry bundle. I wanted a go. 'Oh, you are awfully sleepy . . .' I realised that Jenny had all but forgotten about us. But then she looked up, a little dreamily. 'This one's idea of fun is to sit on my fingers and be stroked, and she just purrs and purrs. I reared her from four days old. She fell off her mother – the mummy carries them for the first week or so, attached to the nipple under her elbow with their curved milk teeth.'

When these cuddly comforters are born, their feet are almost as large as their mothers'. At birth, the mother hangs upside down (or the right way up, depending on your perspective) with her thumbs and catches the newborn in her tail. It then crawls up her fur and latches on to her milk teat. Here the baby is enveloped in its mother's wings, and all you can see is a little bump.

Sometimes the system does not work perfectly, and the baby will drop off during flight. 'Normally this doesn't matter, as they flutter to the ground like a leaf and open their mouths and make a very loud social call,' explained Jenny. 'Which in batty language is, "Mummy, come and rescue me, I am here!" She turns around, flies down and picks it up. This one was not so lucky – flying around a farmhouse in Sussex, it fell off onto the patio and into a crack where Mummy couldn't find her. In the morning, the owner was having his coffee on the patio and he heard "squeak, squeak, squeak" and looked down and saw this pink thing about the size of my thumb joint. Luckily he was no fool and realised it was a baby bat. He called the local bat warden, who came over and looked in all the farm buildings, but could find no sign of the maternity roost. It was brought to my hospital to be hand-reared.'

And then, Jenny explained, a strange thing happened. I'd never really considered bats as particularly social. Sure, they might roost in groups, but how deep could a bat's love go? Jenny put the baby serotine in an incubator on the kitchen counter, along with several baby pipistrelles. And as they grew together they started to smell alike. But the serotine is much bigger, and by the time it had its first coating of fur, the little pips all thought that Mummy had returned and jumped on her.

There is more to bats than you might at first think. And Jenny was about to surprise me again. She reached into another cage and rummaged around before emerging with two very chattery bats. 'Here lies a story,' she began. 'A larger bat and a little one. The larger bat is a Leisler; it covers three fingers across my hand, and the little one is a *Pipistrellus pipistrellus*, covering just two fingers. The Leisler came to me as a juvenile in real trouble. Her mother must have abandoned her and she came in with no nails of any description; her body was lacking in nutrition. Her wings were see-through. She was starving and went into the foetal position, as if she was about to die. I normally massage them at this stage, but it didn't seem to help. When I am really stuck with a traumatised bat, my last option is to try a little bit of pip-therapy. I have found that pipistrelles have a unique ability to help other bats. So I put several in with the Leisler's bat, and they went right up to her and cuddled her and she began to recover from that moment. Within three weeks she had grown her nails and she could climb and groom. But she had become completely insepara- ble from one of her little friends.'

Jenny described the Leisler's reunion with the pipistrelle after it been away for a few days in deeply moving terms. As Jenny was putting these two lovebirds back she carried on talking, almost to herself. 'I love my bats, they make me so happy . . .' She turned back to us, with a beatific smile, and said, 'I have been saving the best until last.' Huma was defi- nitely quivering now. 'I think this is my favourite animal in the world,' she managed to say.

Another rummage and a very sleepy-looking thing emerged in Jenny's hand. It looked like a cartoon character, with improbable ears flopped over its face. Huma was doing small jumps and saying, 'Just look at the nose! See those freckles? He's just so cute.'

Like time-lapse footage of a leaf unfurling, the ears slowly began to take shape, revealing the face of a brown long-eared bat. The ears were remarkable. There is a very small space between cute and ridiculous, and somehow this bat managed to occupy it with confidence. And there were freckles.

The shape of the bat's ear intrigued me; there was a small flap of skin in the front that looked a little like a lid that would shut the ear if the noise was too great. But I was quickly corrected by Huma. 'That is the tragus. All bats have a tragus; it's a tiny ear at the base of the main ear, an extension of the inner lobe. It's a receiver for echolocation – the main ears rotate to guide the sound onto the little ear.'

Evolution really accelerated around 530 million years ago, during the so-called Cambrian explosion, when one thing started to try and eat another thing. This encouraged the eaten to evolve ways of avoiding being eaten and the predator to develop ways of predating. This arms race continues and is elegantly illustrated by moths and bats. Bats use echolocation to find moths. Moths learn to listen out for the echolocation and take evasive action. The brown long-eared bat whispers through its nose the faintest echolocation, and also has hearing so acute that it can, apparently, hear a moth walking on a leaf.

And the way that this freckly bat caught its prey was also pretty sophisticated. Jenny opened out its wing. The hook on what looked like its shoulder was actually its thumb. Only when the wing is held open do you get an idea of what is really going on. 'The wing is the equivalent of the little flap of skin between your fingers, if it had grown to fill the whole space between them. Bats fly with a large webbed hand. In fact, bats belong to the order Chiroptera, which comes from two Greek words: *cheir*, "hand", and *pteron*, "wing". The wing membrane is two layers of skin with blood vessels and tiny muscles in between. It goes round the body, between the fingers and over the tail. This is vital to bats – it's a rudder that enables them to twist and turn in all directions. By opening and widening their back legs, the brown long-eareds can form a little bag, using it a bit like an ice cream scoop in which to scoop insects up off leaves. Then they do a half-somersault to grab them with their teeth. They are the most sophisticated of all our gleaner bats. They have developed methods of catching insects that other gleaners haven't even dreamed about.'

I've had a picture of this very bat on my computer desktop for a while now. There's no denying that it is divertingly cute. Nor is there any denying that bats are fascinating. Or that they are in need of our love and attention. But do they have the capacity to win over our hearts? Huma and Jenny are besotted. Could I become besotted?

As we travelled back to London, Huma said, 'Did you know that the short-nosed fruit bat from China is the only other animal on the planet that has been found to indulge in fellatio? It seems to be something to do with prolonging the more conventional approach.'

Being the slightly cautious person that I am, I made a note of this and decided to check it out. After all, I myself have occasionally been guilty of telling tall tales to the gullible. Delightfully, Huma was not making this up. A fully peer-reviewed academic paper, with the title 'Fellatio by Fruit Bats Prolongs Copulation Time', was published by the Public Library of Science in the journal *PLoS ONE* in 2009.

Huma was already plotting another adventure. 'I'll take you to the hibernation roosts around Highgate, then you'll get to see them in the wild, but close up. Just like you can with a hedgehog.' She was very tuned in to the competitive element of this story, and she's such a force of nature that I quailed at the thought of not choosing the bat. But would a brown long-eared bat look ridiculous tattooed on my leg?

4. DRAGONFLIES

'*Aquatic alien, aerial predator*'

Dragonflies are one of the most beautiful animals on the planet; and also one of the most hideously ugly. Never mind aesthetically challenged ducklings, this is an animal that mutates from science-fiction horror beast into symmetrical beauty, discarding its skin in an act of resurrection. I once watched a mini-swarm of southern hawkers erupt from iris leaves in my neighbour's pond. Erupt may be a little fanciful, it was quite a slow process. Somehow a group of six-foot-tall supermodels were managing to extricate themselves from a fleet of impossibly small cars.

How much do my ambassadors resemble their animals? I was expecting my dragonfly lady, the delightfully named Ingrid Twissell, to be, at the very least, slightly eccentric. But before I could find out, I needed to get the rest of the family settled. My wife, Zoë, and our two children, Mati and Pip, were taking advantage of my dragonfly expedition to tag along to the Cotswold Water Park and enjoy a session in the water fun-park attached to the nature reserve I was aiming for. It was the day after my birthday and I was a little under the weather.

If there was a volume control on life, the water park would

be bearable. But it is just off the main route between Swindon and Cheltenham. And there is no volume control. So despite this being a Sunday, the noise from the traffic was intrusive. But there was a strange sense of artificiality about the place. At the same time, though, there were obvious efforts to inject something real, something natural. In fact it all felt a little like Milton Keynes: new, evolving and full of potential.

After depositing my monsters, I went back to wait for the (I hoped) eccentric Twissell. While waiting, with a cup of coffee in a cafe near a large lake, I read through a leaflet on the history of the park. And this helped explain why it felt like a 'work in progress'; because the Cotswold Water Park *is* very much a work in progress. The gravel extraction that has created the 150-or-so lakes is ongoing; there are two million tons of gravel dug from this catchment area of the Upper Thames each year.

What has taken millions of years to accumulate, through the erosion of the Cotswolds by the action of the river, is being removed at a remarkable rate. Digging began fifty years ago and the resource will be exhausted in another fifty. I wondered if I was alone in wondering whether this was necessarily such a bright way of doing things.

Ingrid arrived showing no obvious signs of eccentricity; in fact, she seemed worryingly normal. We had not had long to talk on the phone when arranging this meeting, so I spent a little time explaining that her role was to try and seduce me away from the hedgehog and that, if she was successful, she would be responsible for determining the subject of my second, and last, tattoo. By now I got the impression that it was *she* who was trying to assess just how eccentric *I* was.

We were settled at a table overlooking the lake when she said she had brought some things to show me, and proceeded to open a briefcase and remove trays of sample tubes. This was more like it.

Each tube contained what looked like an extra from Ridley Scott's film, *Alien*. There is absolutely no doubt where he got the inspiration for the impossibly monstrous creation. These are the larval cases of the beasts we seek, and I am sure if I were to bring one of them on board a spacecraft, there would be considerable unease. How could something quite so hideous turn into something so graceful?

'Well, the *how*, if we are being practical about this, is the same process by which caterpillars turn into butterflies – metamorphosis.'

Ingrid Twissell was clearly going to need some warming up.

As we peered into the sample bottles, holding them up to the light, I was relieved to see that we were getting strange looks from the other patrons. The specimens were remarkable. Ingrid has, over the last thirty-or-so years, been an avid amateur. She has collected these dragonfly memories as part of her way of exciting people into a love of this extraordinary and ancient insect.

'This exuvia,' she said holding up a two-centimetre-long, almost translucent specimen, 'contained a common hawker.' The shape of the larval case looks just like a wingless dragonfly. Because, while these animals do metamorphose, they do so in a different manner to, say, butterflies. But before I can get drawn into a detailed discussion about the lifecycle of the dragonfly, Ingrid started to tell me about where this particular specimen came from.

'I was out on a bug-hunting walk with a group of enthusiasts,' she said. 'We had noticed this dragonfly about to emerge. It was completely out of the water on a thick reed, and a sharp-eyed friend had noticed the cuticle had split. So we stopped and watched, totally engrossed by this moment of near miracle when an old skin is shed and a beautiful animal emerges. Slowly, slowly, the head of the insect pushes down and out of the back, before rearing up perpendicular to the old skin. When it has pulled itself out it rests and lets the rest of the magic take place, as the wings unfurl and the abdomen extends. All in all we were watching this for well over an hour. And then, the moment of the first flight. By now I was feeling really quite involved, this was "my" dragonfly. Take-off was without warning, the engines of the wings turned on and it lifted off with such grace and ease – only to be hit by an emperor dragonfly within a very few seconds, dismembered and eaten. So I kept the case, the exuvia, as a reminder of the reality of life as a dragonfly.'

And that reality is one of predation. From the earliest larval stages right up to iridescent adulthood, these are some of the most perfect predators on the planet. It is no wonder that the name of the taxonomic order to which they belong is *Odonata*, which translates as 'toothed ones'.

The larvae, who lurk in still or slow-moving water, are equipped with a remarkably adapted lower lip, the labium, which can be explosively extended to catch passing prey. This is where the Alien's modus operandi sprang from, I am sure.

Though the vast majority of a dragonfly's existence is spent in water, it is for their brief life in the air that they are best known. Is there a more distinctive creature on the planet?

Certainly there are few as brilliantly adapted to life as an aerial predator. Praise may be heaped on the peregrine for its stoop, or the golden eagle for its size. But compared to the dragonfly, they are both clumsy. Perhaps some bats and flycatchers have an agility that is in the same league. But the consummate ease with which the dragonfly commands its element is magisterial.

Their ability to hover, to fly sideways and backwards, to stop and start on a pin – this aerial virtuosity is all part of what makes them such a deadly threat to anything smaller than themselves. And that would be bad enough for potential prey. But then you consider the amazing eyes. Compound eyes, composed of thousands of small lenses, make up most of the insect's head and allow it to resolve the world to amazingly fine detail, capturing movement in an instant. To cap it all, there is a more subtle adaptation that gives this animal such an advantage. When you next get a chance to watch a dragonfly, look closely at its head. See how it remains amazingly still while the wings clatter and whirr. This is thanks to tiny, hair-like structures at the back of the head that were not noted by science until 1950, which act as shock absorbers.

Before meeting Ingrid I had tracked down the dragonfly bible: Philip Corbet and Stephen Brooks's contribution to the ever-amazing Collins New Naturalist series, first published in 1960. And while it is a wonderfully detailed account, it left me with questions, not least of which was why did they refuse to write about the animals using English, colloquial names?

'The naming of things is a complex art,' Ingrid explained. 'Even with the apparently simple definition of what a dragonfly actually *is*, there are arguments. And this is partly because research into these animals is relatively new. They look amazing on the wing, but they do not keep well as specimens; they rapidly become drab. So there was no great craze of collection and cataloguing among the Victorians; certainly nothing on the scale that was invested in moths and butterflies. In fact it was only in 1977 that the first proper guide in the form of a key appeared.'

The importance of the arrival of a suitable guidebook is easy to underestimate. But as with Ivan Wright and the solitary bees, so with Ingrid; the book gave focus to her love of the natural world – allowed her, and many others, to look more closely at the Odonata.

'You think of the RSPB, which has been around since 1904; but the British Dragonfly Society was only formed in 1983. And I've been a member since the very first day.' Ingrid is now a dragonfly official: she is the County Recorder for Gloucestershire.

The national Recorder scheme is one of the wonders of the world. Passionate amateurs volunteer to collect and collate data for a group of animals or plants and then feed that data to the appropriate national schemes and databases, such as the Biological Records Centre and the National Biodiversity Network. The original citizen scientists, the Victorian enthusiasts, did much to propel the study of nature from a hobby to a serious endeavour. And now it is important that we do not leave such an important endeavour entirely to those for whom it is a profession. Data-collection is something we can all join in with, and frequently do. Events such as the RSPB's Big Garden Birdwatch might be the springboard someone has been looking for, leading them into taking a more active role, perhaps even to become a local recorder for the species to which they are most attracted.

'Dragonflies are one of the easiest animals to get involved with,' Ingrid argues. 'Not only are they stunningly beautiful, but there are also only forty-two resident species, so there's not too much to have to learn.' In fact Britain has a very impoverished dragonfly fauna, so there was excitement when the beautifully named willow emerald damselfly started to breed. There are around 120 species in continental Europe, but that is dwarfed by the full extent of Odonata diversity. In total some 6,000 species have been described, with new ones being found at a fairly steady rate, on every continent except Antarctica, and from the equator to the tree line – the point up a mountain above which trees can no longer grow. Inevitably, the pressures we are applying on the most diverse habitats, particularly in the tropics, means that many species will be lost before we have even discovered them.

It is a shame that a group of animals that were among the first to make their life on land should be suffering such depredations. They have been around in pretty much the same form since the Upper Carboniferous period, more than 300 million years ago. 'Some of the biggest fossil dragonflies have been uncovered in Britain,' said Ingrid. 'Perhaps the most famous was the Bolsover Beast, though this was not the biggest. It was found in a coal seam in Derbyshire and was about the size of a seagull.'

In fact the biggest of all had a wingspan in excess of 70 cm. Which dredged up a question from my A-Level Biology lessons: how *could* a dragonfly be that large? I remember being taught that there was a fundamental limit to how big insects – and in particular flying insects – could get. They do not breathe as we do, through the inflation of lungs. Instead, oxygen for metabolism is inspired through so-called spiracles in the sides of the animal, and while there can be some muscular pumping, it is a much more gentle process than the one that occurs in mammals. The problem is that there is no way for them to get enough oxygen into their system to sustain a body of any significant size.

'The most reasonable explanation,' Ingrid told me, 'seems to be based around the different composition of the atmosphere 300 million years ago. Then there was around 50 per cent more oxygen, meaning that the insects were able to grow larger.' And this is not just theory. Scientists at Arizona State University have succeeded in growing dragonflies up to 15 per cent bigger by keeping them in levels of atmospheric oxygen similar to that found in the heyday of the prehistoric dragonfly.

'But you shouldn't get hung up about size,' Ingrid continued. 'The ones we have today are utterly remarkable enough for most people. So let's go and see what we can find.'

I looked at my watch; we had already been chatting for an hour and had not left the cafe. So we headed off to the gravel pit we had been ignoring while we were rummaging through her samples. It was unprepossessing, and not the sort of wild place that looked likely to contain much that was wild. Not too far away, my wife and children were frolicking in armbands, and there was a network of busy roads right behind us.

'But without the roads, there would not be the habitat, and without the money generated by the water park, there would be less to invest in maintaining this important wildlife reserve,' Ingrid explained.

I remained unconvinced. As we walked, we talked. Apparently we did not need to worry about frightening the insects off with our voices.

'You were asking about breathing,' Ingrid continued. 'Well, I find it always amuses the groups I take on field trips when I tell them how the larvae breathe. Through their bottoms! The dragonflies also use their diaphragm to suck in a great gout of water through their tails and then expel it to make the larvae move by jet propulsion. If they get caught, some shed their caudal lamellae, at the end of their tails – like a lizard shedding its tail.'

I knew that this was just the sort of thing that would appeal to children – the perfect material to excite budding naturalists. As were the dragonflies' names. If animals are going to stand a chance in my rather arbitrary scoring system, one of the most important characteristics will be an ability to enchant. Part of that enchantment comes through the immediate aesthetic appeal of the species, but a large part comes from the stories that surround it. At first glance, toads might not stand a chance in competition with an otter; but the stories of toads are just as magical, as we shall see later. Dragonflies, on the other hand, seem to be doing pretty well on both counts.

Our presumptions about the animals are revealed in the names we give them. In fact, our instinctive reactions to a large, regal insect are understandably tinged with fear. We have learned, over thousands of generations, a healthy respect for brightly coloured buzzing beasts, and the folklore surrounding these miniature dragons is clear.

Adder-bolt and adder-fly are two of many names that link the stingless insects with venomous snakes. These develop a little more with horse-stang and hoss stinger, from which you can begin to understand the misunderstanding. People trust the instinctive intelligence of their horse and then see its reaction to a buzzing insect and make the connection. If these small, stiff, short-bodied flying snakes can hurt a thick-skinned horse, what are they going to do to me?

Couple that with the motions they make when getting ready to lay eggs, and you have a ready-made monster. Sometimes a dragonfly will alight on an object –an arm, for instance – and test its suitability for ovipositioning (egg-laying) by gently prodding the surface with its tail. This requires an arching of the abdomen and could easily be mistaken for an attempt to sting. The stinging myth was used to coerce children, with one old rhyme contending that the 'snakestanger' would show good boys where the fish hid, but would sting those who were naughty.

If I could inspire children with a love of dragonflies through bottom-breathing and stinging, then the next story Ingrid told me had enough horror to excite even the most jaded. Where on earth did the name 'devil's darning-needle' come from?

'You've seen the way the darters, in particular, will fly to and fro across the water?' Ingrid said. 'Well, this was taken as an action of darning. There was already an association with the devil. The body of the dragonfly could be taken as snake-like or resembling a broomstick, the method of travel of the devil's emissaries. Couple this with the darning flight pattern and you have a great story to intimidate children into obedience.'

And this is how it went. If you told lies or used foul language, the devil's darning needle would come and sew your mouth closed. Or, in another version, sew your ears closed, so that you would not be able to hear the good word of the Lord.

As we paused in the first of the convenient bays the developers had left around the twelve-hectare lake, there was a small swarm of delicate damselflies. I was ready to move on – we weren't here to see damselflies – but Ingrid wanted to correct my taxonomic oversight.

'Damselflies are dragonflies. And dragonflies are dragonflies,' said Ingrid. Seeing my confusion, she continued. 'There are two sub-orders of Odonata: Zygoptera and Anisoptera.' Which is basically what she said the first time. Zygoptera are the damselflies, while Anisoptera are what we typically call dragonflies.

'I mentioned earlier the lack of colloquial names,' she added. And it was true – it felt as if Ingrid had been speaking almost entirely in Latin. 'Well, I am not just doing it to show off, you know. It actually makes life much simpler.'

I felt like asking her to pull the other one, but she was already anticipating my protest. 'Overseen by the International Commission for Zoological Nomenclature, the Latin binomial allows each species to have a name that is recognised all over the world. We are then free to call it what we like colloquially. You know that there are lots of different kinds of robin around the world, don't you? Someone tells you they've seen a robin; exactly what species they are referring to depends on where they are. And it's the same with drag-onflies; there have been all sorts of problems with the same species getting different names.'

The New Naturalist guide puts it delightfully: 'When some writers began to show originality in their use of English names, it became desirable to standardise the nomen-clature . . .' In a guidebook to the dragonflies of Ireland, the authors had used names more commonly found on the conti-nent. So what had been known as the demoiselle was listed as 'jewelwing' and the emerald damselfly was down as 'spread-wing'. Of course, this was far from ideal and illustrated perfectly the advantage of the less poetic Latin names.

'So what's that?' I asked, as something brown shot by the corner of my eye. 'A sparrow,' Ingrid said.

We came to a patch that was sheltered from breeze and noise and full of lazy late-summer haze. 'You cannot go look-ing for dragonflies in the cold and wet,' said Ingrid. This was just the same argument that Ivan Wright had used about soli-tary bees. It seemed that UK mammal ecologists had chosen their species quite poorly. And dragonfly hunting seemed even more laid-back than Ivan's athletic lunges for passing bees. What do you actually do when you are looking for dragonflies?

'You have to do a little homework. We know when they are likely to emerge, different ones at different times of the year. And we know that they are going to emerge from particular habitats: some like chalky water, others prefer peaty water; some like ponds, some like lakes,' Ingrid explained, and reminded me of another winning fact: 'There are only forty-two species to remember.'

So how do we tell one species from the next? Should we be here with nets, ready to leap into the water? 'If you've done your homework, then the odds are definitely in your favour, as only a few species are likely to be present in a given habitat. Then it is largely down to jizz.'

Jizz, I remembered, was the term Ivan had used in reference to bees, and birders use about birds – that semi-mystical ability to immediately recognise a species. I have developed something like a 'jizz' sense about hedgehogs, and am able to sense their presence from only a few clues. But I could not compete with Ingrid's ability to glance at a passing buzz and instantly declare that it was a migrant hawker.

'The flight was quite jerky,' she said, trying to fill in the gaps from my glimpse of a blur. 'They emerge in late July, they like standing water and the shallows here are ideal. Now, if we can get closer . . .' It had paused on the stem of a reed emerging from the water. 'You see, it looks quite blue, so it's a male. The females are browner. If we can get even closer you might just be able to make out a golf-tee shape on the upper part of the abdomen.' I peered; it flew. That is why the jizz is so useful; sometimes there simply isn't the time to peer at leisure.

But then I noticed something else on a stick poking out of the water: a dragonfly with its back arched. Ingrid declared that it was an ovipositing brown hawker. Sitting on the twig just an inch above the water, she was probing the rotting branch and inserting hundreds of small oval eggs into the bark, just below the surface, one at a time.

'One of the fascinating things about the eggs being laid just there is that they will enter *diapause* in a couple of weeks,'

Ingrid said with some excitement. I was at a loss. 'They will develop like normal,' she explained, 'but then, in anticipation of winter, their development will be suspended until the water is warm enough again to allow the larvae to flourish. And look,' she exclaimed, 'there are more of them! There are three; I've never seen three of these all laying eggs at the same time.'

And there they were, a synchronous ballet of dipping dragonflies preparing the lake for the next generation of iridescent joy. As we stood up from our intrusion into their intimate world, there was a buzz above Ingrid's white sun hat. This dragonfly proved a challenge to identify as it was obsessed with her head and therefore I was required to relay the description to the expert.

Each of the four wings had a dark spot on the leading edge; the fine body was banded with segments of orangey red separated by pale bands; where the wings attached to the thorax, at what could be construed as the shoulders, there was a yellow patch. By this time the animal had settled on the rim of Ingrid's hat, which she removed before declaring the specimen to be a male common darter. 'You see that yellow stripe down its legs? That is one of the key diagnostics. These are one of the last dragonflies to be found on the wing in Britain each year; sometimes they'll still be flying in November.' And she popped the hat back on her head, where the darter remained like an impossibly brilliant brooch.

This was the nearest I got to a dragonfly, and it was a remarkable thing, to be so close to something so ancient and so wild. But it also seemed alien. And this is what made me feel it was not the perfect vehicle for spreading the message of love that is at the heart of my quest. I needed people to be able to relate to the animal I chose, to be able to look into its eyes and feel some sort of kinship.

As we parted and I headed off to try and reclaim my family, I found myself becoming quite absorbed by what I was trying to achieve. In many ways the dragonfly is a brilliant tool for communicating the wonder of the natural world. It is both hideous and beautiful, it is loaded with myth and legend and its vulnerability to disturbance and pollution makes it a great indicator of the state of our water. It would also make a magnificent tattoo.

But the dragonfly lacks humanity; it is difficult to anthropomorphise a dragonfly. The essence of my quest, I suppose, is empathy: we need to be able to 'get' the selected animal, and the dragonfly's beauty is too strange.

Steadily making my way around the large lake, I eventually came upon my gleeful family. But I was distracted, and found it impossible to get excited by the splashing fun. I wanted some time to think.

5. HOUSE SPARROWS

'Little Brown Job'

The house sparrow is as undistinguished a bird as it is possible to imagine, a species regarded by many as little more than an avian mouse. Even Alfred Newton, the founder of the British Ornithologists' Union, remarked in his 1896 *Dictionary of Birds* that the house sparrow was 'far too well-known to need any description of its appearance or habits'.

And he has a point. It is the epitome of an 'LBJ': dull, drab, plentiful and boring. How can I get excited by this Little Brown Job? How can the house sparrow win a place in my heart (and therefore on my leg)?

Well if anyone can sing the praises of this chattering bird, it is Denis Summers-Smith. In the course of our time together I became progressively more amazed at what this man had achieved in his life. I also began to feel rather inadequate.

For example, Denis began by talking about an extensive period he spent convalescing, which allowed him the opportunity to develop his love of birds. 'I did have rather an eventful war,' he concluded. And I tried to do the maths, before blurting out, 'So how old are you then?' Amazingly, Denis was about to turn ninety. If I have a fraction of his passion and energy at sixty I will be impressed.

It seems a little incongruous that someone who commanded troops on D-Day, before being injured by a shell, should become the world's leading expert on *Passer domesticus*, the house sparrow. So before we got onto the biology, physiology, behaviour and ecology of the bird, I wanted to find out how this happened. And that required a little history.

Very sweetly, Denis had warned me that he was not very good in interview, and would need prompting along the way. So I prepared myself for active engagement. I need not have worried. After a few minutes I realised that I was largely redundant as an interlocutor and settled down in his Teeside bungalow to enjoy the ride.

Childhood holidays in Donegal with a strange uncle who was both naturalist and minister inculcated in Denis a love of wildlife. 'I can still identify all of the flowers that bloom in Donegal in July and August,' he said.

'My father had been a solicitor and my elder brother was one too. I determined from an early age that I did not want to follow this path of dusty filing boxes; somehow I knew I wanted to be a scientist.'

The war intervened. It is a fascinating reminder of how lucky I am. Denis was commanding men in his early twenties; I was studying hedgehogs. He talked about the madness of war. Soon after D-Day, near Bayeux, he had ordered his platoon to dig in on a hill between their positions and the Germans in the nearby town. This was to be done on a very dark night. Strict instructions: no smoking, no talking. He went to check on progress, to discover that the Germans were doing just the same thing, and trenches by both sides were being dug silently and adjacently.

The explosion that took him to ornithological convalescence nearly killed him. And his experience at the cutting edge of penicillin left him with a lifelong fear of needles. 'It was like golden syrup, and I got an injection every three hours for five weeks.' But then he was given freedom to recover in South Ayrshire, where he met a kindred spirit with whom he spent the days watching the birds.

Recovery let the real world back in, and Denis found himself finishing his degree, getting a job, a PhD and a family. Though he has never had a copy of his thesis. The bigwigs at the Atomic Energy Research Establishment at Harwell deemed his research into the physical metallurgy of uranium alloys a little too sensitive and kept hold of his hard work.

It was around this time that he made a decision that would colour the rest of his life. 'I wanted to study birds in my spare time and I didn't want to just watch birds; I wanted to learn about a species in depth. We were living in the Hampshire village of Highclere and it would have been great to head up onto the downs to study the stone curlew. But I had a small child, and this was a time of petrol rationing. So I looked around me for something near to hand, and what is nearer to hand than the house sparrow?'

So this wasn't a Damascene moment? 'I tend to make controlled and deliberate choices; it was just an accident that sparrows happened to be on my house, but I made use of that accident.'

And how he has: while work and family commitments would be enough for most people, Denis has also managed to publish five books and over forty scientific papers on house sparrows. And he's still at it. He showed me the paper he's co-authored with a Japanese friend. 'I'm particularly excited, as he is another amateur, a director of Brother, the printer company. We were trying to make sense of sightings of house sparrows on the west coast of Japan in 1990.'

House sparrows emerged from the Middle East and then followed the spread of agriculture throughout Europe, Asia and North Africa. Since the mid-nineteenth century their spread has been accelerated as people have moved further and faster, and the species can now be found over most of the inhabited world. In fact it is the most widely distributed wild bird on the planet.

But they did not make it to Japan until 1990, when they were seen in seven different places. These were initially thought to be stowaways escaped from a boat. However, the pattern of their arrival did not convince Denis, so he carefully examined all the records, and has concluded that there is only one way this could have happened. A freak weather event.

And this is what his new paper argues: that a storm picked up a flock of house sparrows and sprinkled them along the west coast of Japan. They have not been seen again, so have either died or interbred with the resident tree sparrows.

The distribution of the house sparrow across Europe and Asia is another area of study that has fascinated Denis. By going through historical documents he has been able to map the species' spread across Russia, showing how it follows the path of the trans-Siberian railway. Strangely, these amazingly adaptable birds have taken rather a longer and more north-erly route to Khabarovsk, though have failed to make it to Vladivostok at the eastern edge of the great country.

House sparrows have failed to make headway in Burma

where, as in China, the main sparrow is the tree sparrow. Now, we know the tree sparrow in the UK – smaller than the house sparrow, it is also shyer of man and rarer. It tends to be a rural bird, living in hedgerows and woodland edges. Which does *not*, however, make it a hedge sparrow: the bird we often call 'hedge sparrow' is not a sparrow at all, but the dunnock, which is a species of passerine known as an accentor, whereas sparrows are more closely related to weaverbirds.

The house sparrow moved east, following wheat; the tree sparrow moved west, following rice. It is delightful to be able to see the path of civilisation being illustrated by the progress of this inconspicuous little bird.

The tree sparrow in Asia occupies the same niche as the house sparrow elsewhere, moving in wherever people settle. It is this ability to tolerate so much contact with people that has allowed the house sparrow to have such a wide geographical spread. And it is still spreading – it may not have settled in Japan yet, but quite recently the species took up residence in Iceland.

Obviously the cold is not a major impediment to their dispersal; after all, they managed to cope with Siberia. Denis visited Tierra del Fuego, principally for his work, but with his heart set on seeing the sparrows. They had been introduced to South America, spreading south from Argentina, and had now made it to the very end of the continent. But the snow was horizontal and Denis did not have much hope. 'The main roads had nearly a foot of compacted snow on them, but the centrally heated houses were not well insulated, so there was a narrow gap along the edge of the pavement where the snow had melted, and that is where I found my sparrows.'

Denis is the first to admit that he has been rather fortunate, but being brilliant helps to generate luck, and so does passion. His work on sparrows was threatening to overshadow his 'real' career. 'I was offered a job at Liverpool University, but I thought to myself, I'm a scientist, but not a very good scientist. I'm not very good with people.'

Harold Wilson's 'white heat of the technological revolution' was busy cooking away and Denis landed a highly prestigious job with ICI. 'My first wife, Margaret, said she did not want to live in Middlesbrough, where the job would be, so I went back to them and asked for double the money, which they accepted. And Guisborough is not too bad really. There were plenty of sparrows too.'

Denis has been very consistent in his life. Two wives, both called Margaret, and both physiotherapists.

The job offered Denis the freedom to roam the world, sorting out the lubrication needs of factories far and wide. He brought with him to this engineering work the same attitude that he had developed with his sparrows, a belief in observation rather than interference.

'It is remarkable to think that the scientific observation of nature is a relatively new phenomenon. Up until the Second World War, ornithology was almost entirely a museum activity, examining skin specimens that had been collected. I was lucky enough to get to know a remarkable man, David Lack, who was then the Director of the Edward Grey Institute at Oxford University. He was one of the first scientific field ornithologists and was immensely supportive of my work.

'The only interference I allowed was to catch the sparrows and put coloured rings on their legs,' Denis said. 'Now there's a training process to go through to get a licence to do this, but back then I just wrote to the organiser and got my licence by return of post.'

This enabled Denis to identify individuals and, from that, to build up a much clearer understanding of the relationships between the birds. 'But I couldn't watch them like I'm watching you now. I would have to watch them out of the corner of my eye, because I found that their behaviour altered if I watched them face-on.'

He learned some amazing things. 'I found that they paired for life, which is unusual in small birds. Most small birds come together for the breeding season; sea birds and other long-lived species will breed for life. I found one pair of sparrows who bred together for six years in the same nest. They clearly recognised each other. If they were feeding together at a bone I had left outside for them, the male would pay no attention to his mate, but would chase away a strange female.'

I have seen hundreds of sparrows over the years, holding court in the garden hedge or marking out a patch of the urban sprawl. But I have never really paid much attention to them. And this is what is so remarkable about Denis. In 1963, in his groundbreaking monograph, *The House Sparrow*, he wrote,

One of the fascinating discoveries of a study like this is the appreciation of just how much individual variation in appearance and character does exist among the animals of one species – the wary and the curious, the timid and the aggressive, the smart and the untidy are all there to be seen; and the disappearance of a familiar character from his place on the roof top is not without its real feeling of loss.

Some of the key features in my quest to find a suitable animal to adorn my leg are that it is accessible, that it is interesting and that it has individual character. So far the sparrow is doing surprisingly well, in theory at least, even if I am still inclined to think of it as, basically, an avian mouse.

And, like the mouse, the sparrow does have the potential to be problematic. Denis pointed out that when he started work on sparrows there was far more published about the economic impact of the bird than its behaviour and ecology.

A flock of sparrows descending into a crop must be a dispiriting sight for a farmer. Granivorous birds need grain, and it is no mystery why the house sparrow has followed people around the world – even into cities, thanks to the spillage from the nose-bags of horses. Their diet has inevitably brought them into conflict with people. But to what extent? To get an idea, I had a look at Roger Lovegrove's thorough, yet depressing, guide to our war on nature, *Silent Fields*. He makes the point that our relationship with this bird is rather inconsistent: 'On the one hand it is welcomed as a constant companion around our dwellings and has even been transported by emigrants to the farthest corners of the old empire to foster memories of home. Simultaneously, it has been castigated as a pest and killed in almost unbelievable numbers over the past 400 years.'

Beginning in the eighteenth century, the war of attrition against the house sparrow in Britain began in earnest. But the remarkable resilience of this bird meant that however many were killed the numbers never seemed to go down – and we know how many were killed because parishes were paying out bounties. This was good for the parishioners, who received 1d. per dozen; but had little impact on the house sparrow population. This is probably because the killing was most vigorous in the late summer, when there was a surfeit of young. More effective control would have focused on breeding pairs in the spring.

In my last book I looked at attempts to control hedgehogs with the payment of bounties, and suggested that this might have been some sort of welfare system: rather than paying people a dole, they were rewarded for a social contribution, however futile.

How to catch a sparrow? Nests would be emptied of eggs and young. Liming of twigs, causing the birds to become stuck, is a cruel technique still employed by the strange songbird hunters of southern Europe. Nets were certainly used; either to catch birds driven from a roost, or as traps set with a bait of corn.

The fervour continued, despite a reduction in parish payments, and resulted in the appearance of locally organised Sparrow Clubs. Lovegrove reported the following: 'Rudgwick (W Sussex) Sparrow Club members celebrated their annual dinner in 1865 rejoicing in the harvest of 5,313 sparrows' heads that year and awarded first prize to Mr Wooberry for his haul of 1,363 birds.'

You can begin to see the scale of things from that figure. Over 5,000 birds killed in one year in one parish. Lovegrove estimated that between 1700 and 1930, in excess of 100,000,000 house sparrows were killed.

And still the overall population remained unaffected.

The persecution was not restricted to Britain. Between the wars, there was a concerted effort to exterminate sparrows in Germany. But perhaps the more significant campaign in relatively recent history was in China. Chairman Mao did some calculations based on the amount that tree sparrows ate and reckoned that killing a million sparrows would free up food for 60,000 people. So in 1958 he decreed that the sparrow was an enemy and a campaign was mounted to kill as many as possible. The effort was vigorous: 2.8 million sparrows were killed in Shantung Province alone. But the effect of this cull was utterly unexpected.

The next year the rice harvest fell dramatically. What the Great Man had failed to understand was that sparrows also eat insects and feed them to their young. And many of these insects are pests of crops like rice. The campaign to exterminate the sparrow was quietly dropped and, amazingly, three years later the population was back up to where it had been.

Such resilience; the sparrow is almost in the same league as the cockroach when it comes to hardiness. Systematic attempts to eradicate sparrows have failed. Which makes what happened more recently all the more shocking; so shocking that questions were asked in the House of Commons.

'Usually an issue like the sudden decline in a species will be reported by scientists, and then the rest of the world takes notice,' explained Denis. 'But not this time; it was Londoners. They were the first ones to notice something wrong. In the late 1980s there weren't many bird watchers in the cities, but there were people who appreciated the little wildlife they saw.'

I hadn't realised the deep affection people who live in the depths of the bright lights of the capital had for their brief glimpses of wildlife. But Denis reminded me of a photograph on the dust jacket of his wonderful book, *On Sparrows and Man*, showing someone looking remarkably like an elderly George Bernard Shaw surrounded by the sparrows he is feeding in a London park.

So it was the people of London who started to ask questions when they noticed the decline in the numbers of their local sparrows. 'It was startling, really sudden,' said Denis. 'The total house-sparrow population of the UK was estimated at 13 million in the early 1970s, but only 6 million at the turn of the millennium.'

However, this does not tell the whole story. While a 50 per cent population decline is dramatic, the tragedy was not evenly spread. The reason the decline was first noticed in London was that here the decline was in the order of 90 per cent.

There had been marked drops in numbers before – for example, when the motorcar came to replace the horse, resulting in less seed spilled on the streets. But that decline was nothing like this, nor was it as sudden.

'It began in 1989,' Denis sighed, 'and we still do not have a definitive explanation.'

So great was the concern that the *Independent* put up a reward of £5,000 for the first peer-reviewed publication that gave a satisfactory explanation. But neither that, nor the questions in Parliament, have provided the answer.

Many theories have been floated. Sparrowhawks? It is, of course, unlikely that a predator will have any impact on the population of a prey species. One of the central, and perhaps counterintuitive, understandings of ecology is that the prey controls the predator; predators are controlled by the animals they eat. The idea that if there is no prey, there are no predators is explained clearly in a great ecological primer by Paul Colinvaux, *Why Big Fierce Animals Are Rare*. How about cats? I am torn about cats. I have a love for purring fur. But I also have a deep hatred for these smirking predators that have wheedled their way into our lives, using us as protection to mount campaigns of shock and awe on garden wildlife. But cats are not responsible for the sparrow decline.

Disease? There was a remarkable bit of work undertaken at the Zoological Society of London in 2005, again sparked by the observations of amateurs. Greenfinches were dropping dead, their bodies being found near garden bird tables. Within a year the population in central England had dropped by over a third. The cause was identified as the disease trichomonosis, which had jumped species from pigeons to greenfinches, and to a lesser extent chaffinches. The good news is that we can assist the finches by making sure feeding and drinking stations in our gardens are kept clean, and that, if dead birds appear, both food and water are removed to discourage birds from congregating and passing on the disease. But no disease has been identified in house sparrows.

'I do have an idea,' said Denis, 'but I am on the judging panel of the *Independent* competition. And also, well, because I first mentioned it in the mainstream media, peer-reviewed journals, which prefer original research, are less keen to publish it.'

Denis had looked at the problem from a different

perspective. Rather than examining the individual birds and trying to work out what was wrong with them, he looked at the overall environment. What had happened represented such a huge change in fortune for the house sparrow that Denis felt something must have changed in their environment. And only one major Europe-wide environmental change coincided with this decline; and this was one that was generally a great improvement: the arrival of unleaded petrol.

How could this be causing a problem? Denis found research that showed that the lead replacement, methyl-tertiary butyl-ether, broke down, on combustion, into toxins that are implicated in the increase in urban asthma in humans. He wondered whether this accidental by-product might be having an impact – not on the sparrows directly, but on the urban insects eaten by their newly hatched young.

The link between lack of food and sparrow population decline was made by a PhD student in Leicester, Kate Vincent, who was finding complete broods dead in nests. Her research was picked up by others, and the RSPB conducted an experiment in which they gave mealworms to half the sparrow colonies under investigation. 'Lo and behold,' Denis concluded, 'the ones with the supplementary food had more young.'

So was the lack of food caused by unleaded petrol? While there is some evidence that this might be the case, there was another environmental change occurring, albeit more gradually, throughout our urban world. A great deal of fuss has been made about the eradication of wild corners in the countryside – the hedges and woodlands that help create the sublimely attractive mosaic of much of Britain's countryside

– and there is plenty of proof that managing them better increases the biodiversity of the land.

But less fuss has been made about the piecemeal extinction of the last vestiges of wildness in our cities. The erosion of the suburban network of gardens has reduced the vitality of this already moribund space. Less fuss has been made because we are more likely to live next to, or *be*, the person who has patioed or decked their back garden, or built an extension on it. And then there is the wholesale destruction of front gardens to make way for off-road parking. All of this has reduced the habitats available for the food species that the sparrow relies on.

So food was crucial. But the RSPB experiment did not result in an increase in those populations that had fed. 'My speculation is that the ones fledging did so at a lower body weight and were therefore less likely to rear their own young,' explained Denis.

I pressed him again about unleaded petrol. Because, if there is an ecological catastrophe unfolding and no one is doing anything about it, then I need to shout it from the rooftops. 'I think the pollution that resulted from switching to unleaded petrol heralded the start of something bigger, coinciding as it did with a steady increase in traffic as a whole, and a big increase in the amount of diesel vehicles on the road,' Denis continued. 'Diesel engines generate particulates. Horrid things called PM10s and PM2.5s, particles as small as 10 or even 2.5 micrometres in diameter. These travel deep into our lungs, and, I believe, also into the lungs of young sparrows that are already under stress from lack of food.'

So, as with all ecological conundrums, this one seems to

lack a simple answer. And it is possible that another explanation will be found. But for now, it seems that Denis has come up with one of the most convincing, even if, as he admits, it is based, so far, on circumstantial evidence.

But this does not stop him working on the problem. 'I am absolutely convinced that the decline in sparrows is related to the increase in traffic,' he said. 'For example, here, in Guisborough, I've been taking a census of the house sparrows in the same way for sixty years now. I count the occupied nests. Even in the busiest places you can easily find them. There will be birds calling and food being taken in.'

In 1998 Denis's census revealed eleven pairs of house sparrows in a five-hectare circle centred on a road junction. In the autumn of that year, traffic lights were installed, altering the flow of traffic and causing there to be lines of stationary cars and lorries. Since then no house sparrows have bred in this area.

I started to question the obvious assumption: surely this particular incident could be related to a wider population decline. But Denis was already on to that. 'You must remember, I've been surveying a much wider area, and there was no decline in the rest of the population in Guisborough. The only change in the environment was the increase in stationary traffic.'

Might the sparrows have just moved? Denis is there again: 'They are a small and short-lived bird, but the main reason why I don't think they've simply moved is that they are about the most stay-at-home bird you can imagine. Most sparrows live out their lives within a kilometre of where they hatch. They are completely capable of flying further, and there are some sparrows that even migrate from the harsh winters of Afghanistan to the plains of northern India. But ours are among the most sedentary birds in the world.'

As Denis had explained earlier, there are house sparrows all over the world. And his work as an engineer allowed him to travel into many of the remoter sparrow hotspots. In fact, while remaining true to his first love, house sparrows, Denis has not been shy from flirting with other species of sparrow. In fact his most enjoyable book, *In Search of Sparrows*, charts his journey around the world as he attempts to find each of the fifteen species of *Passer* out there.

But there is a problem. All this research by Denis and his colleagues is unmasking previously unknown species. If he was to try to locate all the sparrow species now, there would be twenty-five to find. Obviously this is not because more species have come into existence; simply that what had once been considered a discrete species might now have been proven to comprise a variety of different species. In correspondence with Stephen Jay Gould, Denis learned that the most appropriate phrase for this accumulation is 'twitcher-driven speciation'.

Back in the UK, the future of our sparrows is partly in the hands of a new international organisation, the Working Group on Urban Sparrows. Because this is far from merely a UK phenomenon; census data from Poland, the Ukraine, Belarus and Russia are all showing declines. Even tree sparrows in Japan are suffering.

And this is important. Not just because any loss of biodiversity is a loss to all, but because these sparrows may well be canaries. A canary in a mine acts as a warning, dying while the levels of toxic gas are still low enough to allow the miners to escape. Can the decline in sparrow numbers be telling us something about our urban environment? It is not as if it would be a surprise. Paediatricians have been studying the development of children in cities and found massive rises in asthma, which is also linked to the amount of traffic pollution.

I wondered whether Denis has considered that it might be time to hang up his binoculars and notebook? Not a chance. Before I had arrived, over breakfast, he had been noting the numbers of sparrows in his garden at five-minute intervals. 'I am an obsessive collector of information. I may never use this particular data, but I'm trying to identify at what point the birds stop coming to the garden as winter draws near. There is still food available, but they just stop coming. They take up residence across the main road in another housing estate.'

The amount of information is staggering. Over sixty years Denis has been meticulously charting the lives of house sparrows. And all the information is safely stored in hundreds of notebooks. I asked if I could have a look, and he pulled 2009 down from the bookshelf in his study.

It was all so neat, so precise – and utterly unintelligible. 'You see, over the years, I've developed a code. Makes everything so much simpler.' Not to me it doesn't; but he started to fill in the gaps. Turning to an entry from late-August last year, for example, he pointed to a note about a male house sparrow with a yellow bill. 'That means the sparrow was becoming sexually inactive. Next time you see a male house sparrow, have a look at its beak. If it's black, that means his sexual organs are functional; by the end of the breeding season it has gone back to yellow.'

Denis doesn't stop with the notebooks. He wants this information to be shared as widely as possible. His enormous database is largely digitised now and he also collects every academic paper published on his beloved sparrows.

'I was asked by one senior ornithologist what I was going to do with all this data when I no longer need it. I could only respond by asking him if he meant when I was dead, as that is the only time I can imagine I would no longer need it. Well, it will all go to the Edward Grey Institute in Oxford. But I circulate as much of it as I can, updating my database and making a CD of it each year, along with instructions, and sending it to anyone who is interested.'

I used the word 'beloved' advisedly. This is a man, after all, who has seen the horrors of war and lived a life rigorously bound to the chemicals behemoth of ICI. When I asked him whether he has to make many concessions to his age, he grinned and said that he had made two very valuable investments. 'An automatic car, because I needed two hands to change gear, and an electric corkscrew.'

Denis is not the sort of person I would imagine emoting about a small brown bird. But when I asked him about his life and what he was most proud of, he unhesitatingly went for sparrows. And it was not all one way; he paused and said, misty-eyed, 'I think, quite seriously, sparrows have kept me alive. They have given me a real interest in life.'

Now that is a pretty powerful way to conclude advocacy recommendation on behalf of the sparrow. But is it enough to make up for the essential anonymity of this ubiquitous bird – or at least once-ubiquitous?

This remarkably dynamic man then offered to drive me to

the station, in his new automatic. As I was getting in he asked whether I would mind if he took his glasses with him. He had been advised it would be better to wear glasses; but he was not used to them. We continued chatting, and as I left the car to say goodbye, I noticed that he had been good to his word. He had brought his glasses with him, but they remained unused, still in their case by the handbrake.

6. WATER VOLES

'A blunt snout emerging from the reeds'

Not all of the animal ambassadors threw themselves at my feet; I did have to hunt as well. And my hunting was often concentrated according to particular requirements. For example, as I looked through my list of potential candidates, the vast majority of them were men.

Some years ago I worked as a researcher for the BBC's Natural History Unit on their radio programmes. One of my jobs was to identify people for interview, and I found that I was nearly always presented with a man. I would have to ask if I wanted to find the equally good women interviewees who were not so frequently thrust into the limelight.

So I was utterly delighted, while rummaging through the wonderful Wild About Britain wildlife web-forum, to see a posting about water voles that had come from 'vole-woman'. This was someone who was, I hoped, both passionate and self-aware.

But how to find her? That was not a simple task. One of her postings gave a link to her blog, 'About a Brook – the diary of a water-vole colony'. This blog included a name – Kate – and a photo. More importantly, it revealed the level of her obsession. It was wonderful. Photographs, hundreds of photographs – of water-vole footprints, water-vole feeding sites, water-vole poo and water voles themselves; all annotated with detailed notes of where and when. But there was no way of contacting her.

I got in touch with the Shropshire Wildlife Trust, as Staggs Brook, Kate's brook, is in Whitchurch. They either did not know or did not want to reveal her details. I started to look through their website and found back issues of their newsletters to download, so I began to do this and search for Kate and water voles. Eventually my stalking paid off and I found a full name, a name that seemed rather familiar. Kate Long.

I popped that into Google and the first result revealed the same photograph as on the brook blog. I thought we might get along when I read the wonderfully absurd quote on the front page: 'In the battle between handbag strap and door handle, far better to knacker your handbag than let the door handle feel it's won.'

And the familiarity? Kate Long is an author, and my wife had just finished her bestseller, *The Bad Mother's Handbook*. It was sitting beside the bed. I wrote to her agent and got a message straight back with a phone number. I called and we talked for an hour. Straight off, I wanted to know whether she was deliberately keeping her two worlds separate. Did she think that a water-vole fetish might affect sales of her perceptive and gently humorous novels?

'It's not a deliberate thing, it's just that these two sides of my life don't often interact,' she explained. 'And I tend to exercise them independently. But the book I am writing at the moment does venture a little into water-vole territory; it is slightly more autobiographical in tone.'

Given the complicated problems she manages to weave into her novels, I wonder how such a blast of the real world will work. And Kate gives me short shrift in response. 'Water voles might be specific to just a few people, but the uniting theme is that every family has problems. I don't think the problems I imagine are any more or less realistic than those experienced by "normal" families all over the world.'

And from there the conversation roamed with the sort of enthusiastic openness one usually reserves for the oldest of friends. We had to stop the phone call after an hour as we both had children to collect from school, agreeing to speak again soon to arrange a time for me to visit her wonderful world of water voles. Since first making contact, Kate has 'come out', as it were, as a water vole enthusiast and indeed her new book, *Before She Was Mine*, features a woman quite like her.

The water vole is, perhaps, better known than you might think. The signature field-sign, the 'plop' as the furry bundle drops into the water of a stream, might be mistaken for another animal, but many people have read about the species without realising it. Who wrote about this animal? None other than Kenneth Grahame who, in 1908, published *The Wind in the Willows*. The second character we meet is Ratty, who introduces Mole to the ways of the river, the two of them boating through a riparian idyll along rivers unsullied by industry. One of the old names for the water vole is the water rat, but this snippet of information sometimes passes readers by who might imagine the character to be a most unexpectedly benevolent individual among that most hated of rodents, the rat.

The confusion is understandable: they are of similar size

and both can live happily in a very watery world. But they do look very different when you get a chance to see them properly, something that the illustrators of the early editions of the book made clear, with Ratty having the snub-nosed features of the water vole. Unfortunately, confusion has been compounded by a more recent edition featuring a distinctly rat-like Ratty.

When I got back in touch with Kate to arrange our fun, everything had changed for her. Simon, her husband, had just had a terrible motorcycle accident and was going to need many operations and many months in hospital. She sounded terribly shaken, but asked me to get back in touch in three months' time to see how her life was then; she was keen to sell the wonders of the water vole to me, but not now.

September came and I found Kate in much better spirits, so a date was set and an adventure planned. We started with food and chat. Simon was in a wheelchair and had a bed downstairs. But he was mending. The accident had been very serious and there had been doubts first about his survival, then about whether he would keep his leg. But now he was looking forward to a time when he could be back on two feet.

While preparing lunch, Kate nonchalantly said, 'The water voles kept me sane, you know. Without them I don't know how I would have coped over the last few months. Even at the bleakest times, I would always find someone to step in for an hour so I could go and sit with them.'

This was intriguing: how could these rather anonymous rodents be so powerful? After lunch we were able to get down to business. Because water voles are quite sensitive to disturbance, we decided to carry out most of our conversation in the comfort of Kate's home. I started at the beginning, with the obvious question of how? How had it all started?

'I was eight years old and travelling with my parents to the West Country for the summer holidays,' she said. 'I remember we pulled into a car park, I think in Crewkerne, to stretch our legs and stock up on supplies. To get into the busy town we had to cross a small bridge, and as we did I noticed an animal sitting beside the stream. I could see it quite clearly, using its dextrous fingers to hold the grass it was eating; and I was hooked. My mum leaned down and said that it was a water vole, but that now we had to get on to do the shopping. And I said no. I'm still surprised by how definite I was, but I didn't want to leave the bridge. Less suspicious days, back in the early 1970s, so my mother eventually agreed to leave me there. So I stood, transfixed, for the best part of an hour. My mother had cautioned me against falling into the water, but what happened was more profound: I did fall. I fell in love. Utterly. My heart was bonded to this little animal and it has stayed with me all my life.'

That is one of the clearest Damascene moments I have come across in my encounters with the animal ambassadors of Britain. And I found it fascinating how Kate was able to openly describe the sensation as love, right from the outset. It has taken me years to get comfortable about publicly admitting to this terribly unscientific way of feeling about a particular species. I think that we lose sight of love at our peril. Because love alone is what will motivate us to care for the world around us. And the lessons we can learn from Kate are vital.

There was a more prosaic impetus to Kate's transformation,

though. She had always had a passion for wildlife – collecting and cataloguing ladybird patterns, and going with her uncle to see badgers, bats or deer. 'But we never saw anything,' she said. 'All that crepuscular creeping through the undergrowth and we always drew a blank. But then that one moment, I *was* able to see a water vole. More than that, I could hear it chewing. And it could sense my presence, and yet calmly it continued.'

This is one of the great attributes of the water vole, argues Kate. Their nature is calm and largely cooperative, if you play the game according to their rules. They will allow you to sit and watch. I said I would ask for my money back if they did not perform during our safari.

'When you go out watching water voles you see so much else – field mice, bank voles, shrews, foxes and rabbits. But none of these animals are happy to hang around; they are all very quick and in a hurry to be somewhere else. But the water vole will sit and feed and swim about in front of you,' Kate said. 'To be honest, I think it probably has quite a lot to do with them being very short-sighted; but I like to imagine that it is just part of their generally lovely nature.'

I can understand this gentle self-delusion. I know that hedgehogs are not particularly desperate to be in contact with me, and that the reason we do manage to establish a kind of connection is more to do with the fact that their defence mechanism – curling up into a prickly ball – does not require them to run away. But that does not stop me imagining that there is some sort of deeper connection. My friend Gordon the Toad would argue that there is a spiritual connection there, but that I am too blind to see. There must be something going on above and beyond the simple absence of flight on the part of the hedgehog; does the hedgehog recognise a benign beast?

Cats, though, would provide an argument against this, as even these domesticated animals fail to recognise revulsion and always make a beeline for the person with the greatest allergy.

Kate's own connection, aged eight, was profound on her part, and has stuck with her ever since. 'Though I did get distracted by life for a while – teaching, family, that sort of thing,' she explained. And not teaching science, which I thought might have been appropriate. 'I've always been more a writer than a scientist, and was teaching English literature, until the books started to flow.

'I had sort of lost touch with my vole-y side, tied up with work, but then we moved here in 1990 and it all reignited,' she said. I had thought she must have moved to Whitchurch *because* of the voles, given how many her blog revealed she had seen. 'I was utterly astonished. Almost immediately after moving here I began to see water voles. I would see them as I carried bags of shopping back from the supermarket; just occasionally, to start with. Anyway, there was a piece in the local paper about a water-vole survey, so I dutifully reported my sightings and got asked if I wanted to be trained how to survey for water voles and, well, the rest is there in my blog.'

That first piece of training really made an impression on Kate. 'With birds, on the whole, it is so easy. You just see and hear them. But mammals are so much more cryptic. You might get a whiff of fox or stumble upon a trundling badger, but it is nothing like birding. And then, after this training, well, as I walked along the canal, I could see loads of signs, all over the place. It was so empowering. Suddenly I was entering the world of the water vole on a different level.'

What sort of signs do they leave, then? I wondered. 'They are really quite generous,' she explained. 'There are five clues to watch out for, of which three are reasonably easy to get to grips with when you have your eye in.'

Footprints, faeces, runways, nests and feeding remains all indicate water-vole activity. 'Water-vole droppings are a good place to start,' Kate said. 'They are often quite visible, as they use regular latrines, and the droppings are about the same size as those of a guinea pig, sort of like brown Tic Tacs. Then you start to spot feeding remains, these are really distinctive. The runways are also easy to find. Look, let's not talk about them, let's go out and see what we can find.'

The air was mild and I asked whether we were going to need Wellington boots. 'Oh, no need for them, we're just going to be walking along where I go pretty much every day. But take this stool, then we can both have a sit down and watch.'

It really was just a short walk from Kate's front door to the end of the town and the beginning of the countryside. Houses gave way to a sports ground and industrial space that drifted on into fields. Hopping over a fence we made our way towards the stream that ran towards the town. 'Now, I can't promise to show you a water vole,' Kate whispered, 'but I guarantee we will find droppings, runways and feeding remains. And I hope that you get an idea of what it is that attracts me back here to my voles, day after day.'

I remember being asked by a wildlife warden on Lundy

Island, when I was around ten, what animals I was hoping to see, and answering 'small mammals'. They have always been at the heart of what interests me. And I am therefore easily pleased by remote sightings, by the evidence of their presence if not their actual presence.

As we neared the stream, Kate pulled what looked like a litter picker from her bag, a long pole with gripping 'fingers' at the end. In fact it *was* a litter picker. She used it to part the long and rich grass in front of us, as a sort of dextrous walking stick. And she was looking worried.

'I think I might have miscalculated a bit,' she said. 'It rained last night, but I didn't think it had rained this much.' As we squelched a little closer to the stream I began to wonder whether I should return to my car to fetch my boots; and then it was too late. I put my weight onto a small island in among the bog, which turned out to be nothing of the sort, and I sank, filling both shoes completely. I grumbled about her footwear advice and noticed that Kate's gentle confidence seemed to be slipping a little. 'I might have been a little hasty about all the field signs,' she admitted. 'These conditions are not brilliant.'

The rain from last night had caused the stream to burst its banks, flooding a whole strip of the field and washing away all the droppings and footprints.

'We will find some evidence,' Kate said, trying to sound in control of things. 'I'm sure. Come on, your feet can't get any wetter.'

And so I struggled under the barbed-wire fence that Kate easily stepped over, and we entered her regular stomping ground. 'What I usually do is pop the stool down just over there and wait,' she said, pointing out a considerable puddle. 'You've seen all the photos on my blog? I took them with a little camera, just sitting quietly and waiting for the water voles to come and reveal themselves. But I'm not giving up today; let's see what's going on under here.'

Kate extracted her litter picker and started to probe the long grass that flourished in the rich moistness of the stream's reach. Parting the stems, she began searching with deep concentration. It took a while, but then she beckoned me over. 'See that?' she asked. 'That is a water-vole runway.'

There was evidence of a sort of tunnel through the vegetation. Kate had advised that the defining feature of a water-vole runway was that you could lay down a Pringles crisps tube in it and it would be a snug fit. And certainly you could. But there I had a doubt. Wouldn't a rat's tunnel be the same size?

'Yes, they would be similar in size, but rats tend to be more nocturnal and therefore need less cover, and so don't make tunnels like this. Now, if we follow this it should lead us to one of three things,' she said. Kate carefully concealed the tunnel as we followed it. 'They are hunted by many things; I don't want to go making their life any harder by leaving their tunnels exposed.

'And here it is,' she said finally, sounding somewhere between triumphant and relieved. 'A feeding station.' Showing the value of her favoured tool she used the litter picker to lift a piece of grass. 'Now,' she said, examining it closely. 'You see this *Juncus*?' Okay, not grass, rush. 'You see how neatly the ends have been cut, at about forty-five degrees? That is absolutely indicative of the water vole. And what they do is collect food, like these rushes, and come to a feeding station and prepare them all, cutting them to roughly the same length, ten centimetres or so. This is just the right size for them to handle easily with their clever paws, and they will make a pile and come back and feed when they get the urge.'

And an amazing variety of plants gets consumed in this way. The voles' convenient storage system means that ecologists have an easy way of assessing what they eat. Perhaps the most thorough of these ecologists is Rob Strachan, who spent two years doing a nationwide survey of water voles and found a grand total of 227 different species of vegetation at their feeding stations. Mainly this consisted of grasses, sedge and reeds, but they are also very keen on nettles, dead-nettle, willow-herb, water cress and, as we shall see later, fruit, in particular apples. Water voles have even been seen collecting potatoes to store in their burrows.

They need to store some food in the burrows, as winter is a tough time to be a water vole. With the cover vegetation reduced, they are more vulnerable to predation, and there is simply less greenery around to eat. So they change their diet and consume more roots, rhizomes and bulbs, pulling them into their burrows from below. They have to be broad in their tastes as they need to eat an awful lot. They can eat up to 80 per cent of their bodyweight in just one day. That is like me eating 65 kg of apples. I'm beginning to develop a new level of respect for these furry bundles.And no wonder Kenneth Grahame had quite such a rich picnic basket for Ratty – though he got the dietary requirements a little off, I am not sure how much of the 'coldtonguecoldhamcoldbeefpickled-gherkinssaladfrenchrollscressandwichespottedmeatginger-beerlemonadesodawater' would have appealed.

Their usually vegetarian diet is supplemented by richer sources of protein, as Kate demonstrated by nimbly plucking a snail shell from the feeding station. And then another. 'You see, they're both eaten in just the same way, quite delicately, all from one side. I'm sure this is the work of a water vole.'

Rob Strachan observed a determined effort to increase dietary protein content in the case of a heavily pregnant female water vole. He watched it swim out to collect fallen pussy-willow flowers that were floating downstream. And then he saw that, despite the efforts required to collect the flowers, only the pollen-bearing anthers were consumed.

Squelching further on along the stream, Kate was busy parting the vegetation, looking for further signs. 'I am determined to find you some poo,' she declared.

But try as we might, there was none to be found. The water had washed the latrines away. In the end Kate suggested we stop and wait, so we settled our stools down with a small splash and made ourselves as comfortable as possible. There was a strange intimacy about the moment. We suddenly became very alone, hidden by the reeds and rushes from what little of the world was passing by.

'This is what draws me here, as much as the voles themselves,' Kate started saying in a quiet voice. 'I sink into the landscape, really; dog walkers pass by, oblivious to my presence. And I'm only a short walk from my front door. This field and the brook just absorb me. You know I said that the voles kept me sane while Simon was in hospital; well, this is what I would do. The big sky above my head, the trees and rushes rustling around me and the water gently gurgling, all helped to soothe my soul. People would ask me whether I came here to think of ideas for the book, and I would say yes, but that was a lie. It was easier for people to accept that I was here working, rather than just emptying my mind, because in reality I guess that what I do here is a sort of meditation. But I wouldn't meditate without the voles. They give me the excuse to get out here.' Much has been written about the healing effects of nature; how the psychosis of city life can be leavened by some contact with nature. And it was clear that even this fairly domestic wildness could be as effective as anything more glamorous.

Calmer, and forgetting quite how wet my feet were, I settled into a dreamlike haze, waiting. I asked Kate what they would be up to now. I know they don't hibernate, but do they spend their time collecting stores of food?

'There's that song – "Cotton-eyed Jo": "Where did you come from, where did you go?" That sums up the voles pretty well. They just vanish over the winter, there are no signs, nothing, and then suddenly it's spring and they're there again,' explained Kate. 'Obviously they've hunkered down in their tunnels, but we're still not clear to what extent they enter a state of torpor, or how much they rely upon stores of food, or food they manage to find underground.'

We've become so used to empirical studies explaining things that it's easy to forget that it was not long ago that people used to think that, for example, swallows did pretty much what we think water voles do, digging themselves into the mud of ponds to escape the ravages of winter. Only with ecological research did people uncover the arguably even more remarkable truth of their ability to migrate many thousands of miles.

'Now I think we might as well try and be quiet for a while,' Kate added, without a great deal of confidence in her voice. 'You never know, there might be a vole around.'

She reached into her pocket and removed a small sandwich bag containing chopped-up apples that were very close to being too ripe. Carefully, she used her litter picker to pull out a chunk, before placing it on a patch of short grass just across the stream from us. Despite all the rain, the stream was only a couple of feet wide. 'I used the litter picker to minimise any of my smell on the apple,' she whispered. 'And that patch of grass is a sort of feeding platform; it's been cropped short by voles. Now, let's just see what happens.'

I'm quite used to this game, and, like the young Kate, I'm very used to seeing nothing. So I was really quite surprised to notice the blunt snout of a water vole emerge from the reeds opposite within a minute of the apple being put down. The snout hovered for a few moments before the rest of the body followed it out, and went straight to the apple, where – bless it – it did what water voles are supposed to do. It sat on its hind legs with the apple between its front paws. I looked across at Kate and she looked as surprised as I felt. We both grinned.

I had been prepared to see nothing, and had accepted that the calming experience of quietly focusing on the movement of the water and the rustle of the reeds was evidence enough of the value of water voles. So to be presented with an archetype of a water vole: perfect. I was as excited as I had been when I'd seen warthogs in Tanzania or rhinos in Namibia. This was real evidence of the value of wildness, and proof that the wildness does not only exist out in the wilds.

I lifted my camera and caught Kate's eye. She nodded and

I took a picture. In the near silence of our meditation I was shocked at quite how loud my camera was. So was the vole, and I have a lovely image of a blur of brown to remind myself of the little character who then excelled itself by making the classic 'plop' sound as it entered the water.

The fun was not over, Kate put another piece of apple down and, five minutes later, a different, larger, water vole came and gave another show, which lasted longer, thanks to my managing, this time, to refrain from trying to capture the image on my camera.

After this one departed we both sat back. I hadn't noticed how intent we both had been, utterly focused on the little beast in front of us. And it was only now that I noticed quite how uncomfortable my feet were feeling. A cloud had covered the sun and my toes were beginning to chill. Wiggling them did not generate warmth, just a rather swampy noise.

The legs of the chairs emerged from the mud with a pop and we began to extricate ourselves from the dense corner of water-vole real estate. And as I looked around, I wondered what the voles did when the waters rose. The evidence of their runs and burrows was all easily within reach of a small increase in water level.

'Those burrows – you see that hole there – well, they'll head uphill, away from the water,' Kate explained as we splodged back to dry land. 'The voles will be able to move to dry ground. But water is far from the biggest thing for them to worry about.'

I had been so tied up with everything else that I hadn't asked one of the key questions. Namely, how are the water voles doing?

Life for water voles can be rather hard. They are a favoured snack of many different animals. This is not, on the whole, a problem for the species. But this can go wrong. And for the water vole it did so with alarming consequences. They were hit with a double whammy. The first impact was one that affects much of this nation's fauna: habitat fragmentation.

The importance many people give to flying animals exercises me. Yes, bats, birds and insects are all important; but to measure the value of habitats almost solely on the status of these winged wonders misses a crucial component of the whole. The idea of habitat fragmentation was first brought home to me by the work I did with hedgehogs. They are the perfect species to show how a habitat can be fragmented and degraded – road casualties providing the most visible evidence. Yet fragmentation happens in more subtle ways, and while a large road will have an impact on water voles, a change in land-use management might cause just as much trouble. If a stream has its banks cleared, or cattle trample its edges, or a marshy area gets drained, water-vole habitat is lost. These processes divide up waterways into areas that are good for water voles and those that are not, creating 'islands' where the water vole is vulnerable to piecemeal extinction.

The other problem – and it is one that nearly wiped out the water vole from vast reaches of the country – came in the form of an alien invader. The American mink had been kept in farms throughout Britain for its fur, a staggeringly barbaric industry that generated great luxury. I have to admit to being greatly conflicted about fur. I know it is usually the result of great cruelty. But I can still feel my mother's fur coat as she leaned over to kiss me goodnight when she was off out to dinner with my father, in fact I can still just about remember the perfume too. I would want to stroke the coat, and I still do, if I see fur in a second-hand shop, I will sneak in a furtive stroke. Luckily the only friend I know with vintage fur is accepting of my need to just touch her arm once in a while.

Some of these mink escaped and some were released. While there might have been a few numpties in the animal-rights movement in the very early days who did not think through what they were doing, what is less well known are the rumours of farmers, noticing the decline in the market, jettisoning their unwanted animals, blaming balaclava-clad hooligans and collecting the insurance. This reaction to campaigns against the fur industry is not restricted to the UK; similar stories have come from across Europe: when there is any upsurge in campaigning against fur, a few mink get released and the activists get tarred as idiotic criminals hell-bent on destroying the ecosystem.

The good news is that no one can be releasing mink into the wild now, as in the year 2000 the last farms were closed down. But the mink are out there and contributing to the decline of the water voles. In fact, it is the combination of intensive farming and the mink that has been pushing Ratty to the very brink.

In Britain the problem can be understood by considering the recently changed Latin name. Until recently, the water vole was *Arvicola terrestris*, which sounds determinedly un-aquatic. The name reflected the fact that, across much of its range, from here to eastern Siberia, the 'water' vole is really quite terrestrial. They are creatures (and occasionally pests) of fields. But in Britain they have an almost entirely aquatic and riparian existence, which is why they have been renamed *Arvicola amphibious*. Taxonomists – those who decide on the names of species – are always fiddling. This was the name they originally gave the animal, before it was renamed *terrestris*.

The reason that this is important in the discussion about feral mink is that our water voles were particularly vulnerable to this efficient killer. The intensification of agriculture, bringing it right up to the water's edge, and the canalisation of rivers with reinforced banks to increase the speed of water flow, have both fragmented water-vole habitat. But they have also helped to confine the voles to narrow ribbons of riparian habitat; perfect mink territory, in other words.

The good news is that this is not an insoluble problem. For some time it was thought that the two species would not be able to coexist and that only the extermination of the feral mink would bring the voles back from the brink. However, if the voles are free from their linear bondage and able to enjoy life in a wider range of habitats, they are able to tolerate the presence of feral mink. Additionally, the sorts of environmental improvements that are being forced upon the industrialised countryside to assist other species, such as otters, also benefit the voles, as the otters tend to inhibit the activity of mink.

A lot of fuss was kicked up about the feral mink and their impact on water voles, and while there was and is a serious problem, it is all too easy to get fixated on that one threat and forget that water voles need to have a suitably varied environment and clean water too. If all the water-vole friendly habitats get developed or intensively farmed, it will not matter in the slightest whether there are mink present or not. So how safe were Kate's voles?

'You have to be on the case; I'm always on the lookout for planning notices. And so are the rest of the Whitchurch Community Water Vole Project, and if we find anything that threatens our animals we're on the phone straightaway. I've heard them in the background say, "It's that vole woman again".' I know I should have been asking about the threats to water voles, but I had to ask. 'There are more of you?' I was under the impression that Kate was unusual – in a good way.

'Oh yes, it started in 2006 when there was a particularly good year for voles, and they were seen right up in town. So a small group of us organised a meeting to let people know about the animals and to also see if I could get more people interested in surveying. Ninety people turned up, can you believe it? It was amazing. And since then we've been sharing information and making sure that important fields don't get developed, because there's been not just a fragmentation of the habitat, but a fragmentation of information between developers, ecologists and amateurs. We're trying to stop that happening, because the picture is often bigger than the one that developers and even ecologists might see. The voles need the waterways, obviously, but they also need to be able to move *between* watercourses. This is not just about one colony of voles here in this field, for example; it's about the meta-population that extends out into the countryside. All of it is interlinked. Wildlife corridors are absolutely critical, and so are information corridors.'

Now we were on firmer ground I sat on my chair again and

emptied my shoes of water. 'You're quite aggrieved about that, aren't you?' Kate said, smiling as I poured the water onto the grass. Walking back home she started to mull over the extent of her eccentricity. I was assuring her that, compared to many people I have met over the years, she seems really rather normal. 'No, I knew it had reached an extreme point when, a few weeks ago, I got a phone call from one of my friends,' she said. 'I put mink rafts down, clay pads that reveal footprints. It's a good way to monitor their spread. Anyway, I hadn't been able to get out for a while, so I asked her to keep an eye on things, and she called to say, "I knew you would want to know that your babies were safe". And it hit me, if that's how she saw me, referring to them as my "babies", I had really tipped from being merely enthusiastic to being slightly odd. To be honest, I'm not uncomfortable with that label. Odd can be good, can't it?'

That's certainly an idea with which I'm very happy.

7. MOTHS

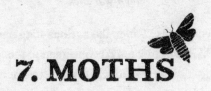

'Extraordinary patterns and colours, beautiful enough to rival the most exotic of birds'

Of all the animals on my list, this was by far the easiest. I had seen more moths, both species and individuals, than any other animal I was investigating. Better still, there were moth obsessives who were trainspotter-ish in their devotion. Often they were bird fans who needed to continue their listing-and-ticking into the night hours. And what was more, the Butterfly Conservation Trust had put me in touch with a young woman who had just completed her apprenticeship in moths, as part of the Natural Talent training programme of the British Trust for Conservation Volunteers. An apprentice? What could go wrong?

Amy Huff was buzzing with enthusiasm and energy. At her flat in Edinburgh we rummaged under her bed to find the first collections of moths she had made. Prizes, pinned and crucified in perpetuity. This has always rather bothered me, the actual collection of specimens. Obviously it would be unacceptable for mammals or birds. And birds' eggs are now protected, even though many revered naturalists cut their teeth collecting the eggs of tits and wrens. So why is it still acceptable, in these marginally more enlightened times, to kill and collect moths? Given the parlous state of our natural world, should this hobby not be consigned to the 'Museum of Embarrassing Things we did when we were Young'?

'It is a bit of a contradiction,' Amy said a little defensively. 'But it is very unlikely we would ever be having an impact on the population of a moth by taking an individual. One of the problems is that there is such great diversity – there are over 2,500 species of moth in the UK. And the differences within a species can be as great as the differences between species, which means that identification can be tricky.'

This was a problem that Ivan, my solitary-bee expert, had encountered. While many of the insects are readily identifiable, most are not, and they need to be examined under a microscope. And it is possible that their individual sacrifice will make a whole lot of difference to the fate of, say, an area of woodland, should the beast in question turn out to be very rare.

But that still did not explain the presence of the cinnabar moth that Amy showed me, gently held by its pin. Even I knew this one. Well, I thought I did. It turned out that this was a garden tiger moth, which is slightly less boldly coloured than the cinnabar moth.

At rest, with its brown and cream forewings folded back, this tiger is pretty. There is little to suggest what lies beneath. But when the back pair of wings flash they give an unmistakable message to potential predators: black-spotted and ruddy-orange, the wings tell of a body stuffed full of toxic alkaloids. The delightfully named 'woolly bear', the caterpillar stage of this moth, also has an armoury of stinging hairs to help keep it safe.

'This is Tony, my garden tiger,' Amy said. 'Of course, there is a contradiction in killing what you're hoping to conserve, but there are some very good reasons for doing so. Firstly, the samples that have been collected over the last 200 years have given us a brilliant insight into changes in habitats. The location of each sample helps paint a picture of shifting environments. And as long as collection is done responsibly, this can continue to help species, as well as identify new ones. But Tony is different. As part of my apprenticeship, I had to learn about rearing caterpillars. This allowed me to watch their change of state into magnificent moths. So Tony was a practice caterpillar, and my first, because they are easy to rear. I collected him in caterpillar form with my mentor, Keith, from a very robust population, as part of a study looking at the parasitic wasps that use them as hosts. After he pupated, well, I couldn't release him out of the window, as I live in the centre of a big city.'

She delicately returned Tony to the collection, all neatly labelled and pinned in three wooden boxes made for her by her father. 'I love these boxes,' she said. 'My dad is amazing, he can pretty much make anything out of anything, and I'm fairly sure that these boxes come from an old chest of drawers we used to have at home.'

The way she talked about her family gave the impression of a pretty idyllic start in life, as long as you are keen on adventure. 'Both my parents were cavers, and they taught me that the muddier and smaller and squeezier it gets, the better it is.' I found myself silently thanking whatever deities were present that I had not chosen a more subterranean species. 'You know, there is a "Huff's Pot" in the Yorkshire Dales . . . well, under them. I think my grandfather found that one.' And seeing me squirm she added, in an attempt to win me over, 'Even on the most beautiful and sunny days, I just love to get underground.'

I don't think of myself as unadventurous, but there is something so deeply primal about my lack of desire to be crushed to a pulp under tonnes of rock. I blame it on my rather square physique; I am just not designed to squeeze into crevices.

I was impressed by Amy's collection, but she dismissed all her hard work with a wave of her hand. 'Really, this is nothing,' she said. 'Just you wait until you see what I have in store for you this morning.'

One of the defining – well, almost defining – features of moths is that they are nocturnal. So where was she taking me in broad daylight? 'Just you wait and see,' she said, grinning as we climbed into her dangerously fragrant car. 'Sorry about the smell, the petrol from my generator spilled and I can't get rid of it.' It was a good job that neither of us smoked; we drove with the windows open.

Edinburgh is rather amazing. I have a great friend who lives out in Leith, near the docks, but this was my first taste of the central scenery. At the end of Amy's road the sky was filled with a mountain. Not far off in the distance, but right there: Arthur's Seat. I'd heard about it, obviously, but seeing how strikingly it dominated the sky was another matter.

Our destination turned out to be a fairly unprepossessing warehouse in the middle of an industrial district. The security guard on the gate knew Amy and waved us into the car park. And as we walked to the unlabelled front door Amy turned to me and said, 'You are about to meet one of the world's greatest experts on moths, and also my mentor. But most of all, you're going to see more moths than you could possibly imagine.'

And with that, she pressed the buzzer and asked for Keith. We were buzzed in and met on the stairs by an elderly but energetic man, Keith Bland. It felt a bit like being introduced to a grandparent. He was obviously very fond, even proud, of Amy and he bounced us upstairs to a staff room, for a cup of tea and – I hoped – an explanation of what was going on here.

I think there was a little conspiracy between them; they seemed amused by my bemused state. 'This is the Natural History Museum's collection,' Amy said. 'This is where all the vast amounts of material that don't make it into the public display cases come to hide.'

So are these inadequate specimens? I wondered, as we drank our tea.

'No, this is the research base,' said Keith a little indignantly. 'Not many people are going to be as excited by thousands of superficially similar micro-moths as I am; and without what we do here we would have no idea of the extent of biodiversity and the threats it faces. Come on,' he concluded, 'let's start downstairs.'

And with that he was off.

We caught up with him on the stairs. I found it rather amusing, given the enthusiasm of these two, that each and every airtight door had a sign on it warning of the risk of – moths.

'There are only two moths we have to worry about, and that is out of a total of around 2,500 species,' Keith said. 'These are the Webbing Cloth Moth [*Tineola bisselliella*] and the Casemaking Cloth Moth [*Tinea pellionella*],which are a serious threat to stuffed specimens in museums. Anyway, everything that arrives here from another museum goes into the deep freeze for four days. That sees to the moths, and the rather destructive museum beetle.' To appreciate the problem, consider clothes moths. The same two species that wreak havoc in your woollen jumpers do the same in a museum filled with the preserved fibres of stuffed animals.

Inside, the large storeroom might superficially have been dismissed as any old storage unit, but already I could see that there were cases in here of great age. It all appealed to my chaotic sense of archiving (you should see my office). A jar of preservative containing a small snake was sitting on top of the vertebrae of a basking shark and what turned out to be the spiral valve from the stomach of a porbeagle shark. These were all sitting on a cabinet dated 1912, which Keith opened, releasing a cloud of the very particular odour of old cabinets. Every time I am hit by that sort of smell I get transported back around 100 years to a room full of Edwardians, all peering closely at recently acquired collections from the far-flung corners of the empire. Somehow the natural history collections feel more honest, less likely to be the result of theft and pillage, but that is probably rather a romantic assumption.

Behind the glass doors was a series of drawers. Beautifully crafted from mahogany, they easily slid out to reveal a static swarm of stunning butterflies.

Which was lovely, but I had come for moths.

'They are all part of the same order of insects, the Lepidoptera,' Keith explained. 'Kingdom: animal; phylum: arthropod; class: insect; order: Lepidoptera.'

A little taxonomy makes the world so much more fascinating. The more I learn, the more I love it.

'Now, the name comes from the Greek,' he continued. '*Lepis*, meaning "scale", and *pteron*, meaning "wing".' This pleased me, as it reminded me that my quest has also been a learning process. Ivan had talked in a similar way about his bees, the hymenoptera.

But what is the difference between a moth and a butterfly? I had been asked the question by my daughter, Mati, a few weeks earlier, when I had corrected her after she called a moth a butterfly. I knew it was a moth, but when it came to saying what, exactly, the difference was I only knew that moths flew at night. But I knew that was inadequate, as I've seen moths out in the day.

So what answer did I get from moth-guru Keith and his apprentice Amy?

'It's very difficult,' they both said, together.

'Basically, butterflies fly by day and moths by night,' Keith said. 'But, as you know, there are lots of day-flying moths and there are even night-flying butterflies, though not in Britain. You have to think of the Lepidoptera as a continuum of families.'

So there's no distinguishing line between them, butterflies and moths? They are not discrete entities? 'The whole of taxonomy is about drawing these lines or boundaries,' Keith said, 'but it's important to remember that they are entirely artificial. There are always going to be species that sit on the fence between families. Actually, we usually just create a new family to stick those ones in, to keep it all neat and tidy.'

There must be more than just day and night to this artificial split, though.

'There are three main rules, all of which can be broken,' continued Keith. 'First is day and night. Second is that butterflies have "clubbed" antennae. Third is that butterflies tend to hold their wings together when they rest, while moths leave them flat over their backs.'

The trouble with trying to find a rule that works is that, if it is to be of any use, it needs to be universal. In Britain this is relatively easy, as there are only sixty species of butterfly. This is not an impossible range to remember, and once you have them in mind, well, everything else is a moth. It may not be the best solution, but it works. If you know the butterflies. Which I keep forgetting. (I have a poster of most of them in the conservatory and gaze at it regularly in the hope that the images and names will become embedded in my brain.) It should not be too difficult. Patrick Barkham wrote a lovely book, *The Butterfly Isles*, in which he found all sixty in a single year. I struggle because some only come out for a few weeks, and then I have eleven months to forget them again.

Back to the drawers. I was intrigued that the specimens were in such a higgledy-piggledy state. I had imagined something far more rigorously organised.

'These still have to be organised and archived,' Keith said.

But they were dated 1912. Had they been sitting here for nearly a hundred years, waiting to be sorted?

'Oh, we didn't get these until the 1950s. The collection had been given to the British Museum but they couldn't cope. This is just one of many cabinets. Anyway, London tried to cherry-pick and then handed us the rest,' Keith said. But I sensed a smile. 'The thing is,' he continued, 'they really are not that good at cherry-picking.'

So when this collection was collated would the job be finished? 'There will always be more; this job will never end,' Keith said gladly. I'm sure that the ancient Greek gods would have considered this an effective punishment, but he looks in his element. What about the bits of shark? 'I picked those up some time ago; they'll end up in their rightful place eventually. Now, let's go and see the main show.'

We meandered back to the storeroom door past stacks and stacks of cases containing the life's work of collectors through the ages. Much of this was waiting for Keith and his team to pick through and catalogue it. The scale of the job was staggering. If someone was considering a specialist career that had a good chance of never running out of work, moth taxonomy seems like a good choice.

'We have some of the best collections in the world,' Keith said, as we made our way back upstairs. 'This is due to a combination of factors. The Empire sent collectors to every corner of the globe. And then there is the simple fact that in Northern Europe the collections last longer than most places, as the environment is less destructive.'

As long as you remember to freeze all of the incomers, I reminded him.

'There was a problem with an ostrich once,' he said. 'We didn't know there was an ostrich to start with; all we knew was that there were moths. We shifted and cleaned everything but they kept coming. Weeks later someone moved a bookcase and found an infested ostrich stuffed against the wall.'

Passing through protective doors, we entered a more sterile

environment. There were no stray bits of shark or snake in here, just rows of cabinets on rollers.

'This is all moths and butterflies,' Keith said with great satisfaction. How many were there? I wondered.

'I have absolutely no idea,' he said. Which felt remiss: surely there was a record somewhere. 'The record is the collection itself; and just think of the hours it would take to put all the names into a database. It would serve no useful purpose other than to answer your question.'

Well, that told me. But I did do a little calculation, and reckoned that in each of the dozen or so rows there were fourteen cabinets. Each cabinet had thirty drawers, and each drawer contained, at a wild guess, 200 moths. Now, do the maths and you get a room with over a million moths – all identified and catalogued.

I could see how in awe of Keith Amy was, and I could understand. He had a completely infectious enthusiasm for these cabinets. He was spinning from one drawer to another, pulling them out, exclaiming over the diversity and beauty. I commented that this was a remarkable achievement, to find a job that was so satisfying.

He looked at me quizzically. 'This is not my job. This is just my hobby. Most of the people here,' he said, pointing back to a room we had walked through full of grey heads leaning over microscopes, 'they are all doing this for love. If it was left to the professionals, this work would never get done. It needs us amateurs. Nearly all of the British insect books – the detailed ones that allow you to identify species with elaborate keys – have been produced by people who were not paid to do so. In fact, I have another one to finish rather soon.'

'Yes, you do,' chimed in Amy. 'I've been waiting for it for ages.'

There was something delightfully incongruous about the contents of this room. The cabinets were all high-tech metal structures on rollers, but the drawers inside represented a leap back to the Victorian collections – hundreds of identical moths, all staked out for us to view.

'These are not identical,' Keith interjected, sounding a little exasperated. 'There are four species in here, and this shows the enormous difference in appearance of members within each species. Here, have a look through this.' And with that he handed me a magnifying glass. It was one of those Lewis Carroll moments; I felt myself tumbling into a different world. I pulled myself back into this one and saw both Keith and Amy looking at me with satisfied grins.

As is so often the case when it comes to observing wildlife, appreciating what was in front of me was a matter of perspective. At a glance, these looked little more interesting than the clothes moths I smear across the bedroom wall in my futile campaign to rid my clothes of them each summer.

'You see that?' Amy smiled. 'Butterflies are all well and good, flaunting their gaudy reds, blues and purples in the sunshine. See this flying by, and you might think it was a fruit fly it's so small, but when you take the time to look closely, you begin to see how beautiful moths are.'

And Amy was right. Through the lens, there was a riot of subtle colour. The wings were patterned with metallic gold and silver stripes, and as I moved my point of view so the wings shimmered.

'Children often "get" moths more easily than adults,' Amy said. 'They're more willing to see the detail.'

'It's not just the flying adults that are amazing,' Keith continued. 'These critters' caterpillars are leaf-miners. They live inside the leaf. You know when you see a leaf that seems to have tracks inside it; they've been made by the little larvae. And it's double win – they're protected from predation and they're also avoiding some of the plant's own defences.'

Pushing that drawer back in and opening another, Keith grinned even more. 'Now, look at this lot, the nemophora, the longhorns. You see the males? Their antennae can be up to six times the length of their body. And see the bodies? If you examine them in natural light, they're like jewels.'

The antennae did look a little silly, especially when I thought of them flying, with more than 2 cm of delicate, wispy sense-organ trailing behind them. Another drawer and another marvel. 'This is *cuprella*, *Adela cuprella* to be precise,' Keith said. It looked so exotic, like burnished copper. Was it tropical?

'I found that one down by a reservoir in Scotland, on a gorse bush,' he said. 'These things are all around us; largely ignored, but there. You don't need to go off on an exotic holiday to be amazed by wildlife. Just try looking a little harder at your own patch.'

And then another drawer . . . I realised that I was going to be in a state of constant wonder as this continued: each of these cabinets held thousands of extraordinary things. This time it was butterflies.

'No, burnet moths,' Amy said. 'They fly in the day and they have clubbed antennae – both butterfly characteristics – but they are moths. When they sit, their wings are not clapped together upright like a butterfly, but flat behind them, like a moth.'

Bright red, these would be predator magnets were it not for the cocktail of poisons the moths accumulate from feeding on ragwort and other toxic plants.

So how did it all start for Keith? Was he a retired professional?

'I'm a vet, mainly research, but I've been passionate about moths for so long,' he said. 'In fact, it started with butterflies, but then I got hooked on micro-moths. My father used to get the bus every day into work, and he'd been talking with someone who worked with one of the foremost moth experts at the time, John Heath. When my interest was mentioned, a meeting was arranged. I would have been about fourteen, and this expert was the person who really opened my eyes to these amazing animals. And he gave me a piece of advice that I have always stuck with; he told me not to bother even looking for English names, just stick with the Latin, they're easier.'

I begged to differ. Especially when the English names of moths are some of the most poetic. How could anyone prefer *Lithomoia solidaignis* over the golden-rod brindle? Or *Anarta myrtilli* over the beautiful yellow underwing?

'But what can you *know* about the moth from the English names?' he protested. 'Not a lot; they're just the hyperbolic outpourings of early collectors. The Latin names put the species in perspective; they immediately tell us what it is related to. And you also have to remember, most moths have no English name anyway.'

It was time for another cup of tea, and also to make plans

for tonight. Because I wanted to see wild moths, and Keith was able to guide us to a spot that he reckoned would be wonderful. 'If you get there while the sun is setting, you'll get a gorgeous view of the ruined Crichton Castle, and there are all sorts of habitats, old-growth deciduous, marsh, a bit of conifer; it's a great site.'

The fumy car was no less fumy as we headed south out of Edinburgh to Crichton. Urban gave way to rural, which in turn descended into old wood. It felt strangely like Somerset as we wove along single lanes overshadowed by tunnels of trees.

Pulling off the road, we unloaded the first bit of kit. 'This is another of my dad's inventions,' Amy said proudly. 'Right out of Wallace and Gromit, this is. It's a self-closing moth trap. The bright light attracts moths, they get closer and closer until they settle in among the eggboxes down there, and when the power goes off as daylight comes, the doors shut, trapping them safely inside.'

We headed off through a field bordered with old sycamore and ash, towards a more open patch that contained a marshy portion, some younger trees and a view, as predicted by Keith, of a distant ruin bathed in golden sun. 'This is perfect,' Amy said, setting down the box and getting to work. 'Look, bats already, and the sun's not even down. This is great, the bats need something to eat and often as not that is moths, so I have a good feeling about tonight.'

So how does this work? Or rather, *why* do mothtraps work?

'Well, what happens if you leave your lights on and a window open in summer?' Amy asked. 'You create a moth trap. They come towards the light, circle it until they get tired and then settle on the curtains. All I'm going to do tonight is replicate that in the field, using egg boxes as opposed to curtains. They're great, so many places for the moths to hide.'

But that still doesn't answer the fundamental question of why moths are drawn to the light. They're nocturnal; surely they should be attracted to the dark?

'There are various theories,' Amy said. 'One that seems highly plausible is that they tend to navigate using the moon. If they keep the moon on one side of them they will tend to go in a straight line. But if an artificial light replaces the moon as the brightest light in the night sky, they will do the same thing, but because the artifical light is so much closer, they end up circling it.'

So how, I wondered, did the passion start for Amy?

'It was a total accident,' she said, as we headed back to the car to wait for the trap to do its work. 'I was working with the RSPB, and was given the job of opening the moth traps first thing in the morning. I was a little unimpressed, first because I would have to be up early, but mainly because I was interested in *birds*, and moths, well, they were just boring. But within a very few days I was utterly hooked. How could I have missed these enormous pink and green elephant hawk moths all of my life? There were extraordinary patterns and colours, beautiful enough to rival the most exotic of birds.'

But what, I wondered, was the point of being bright green and pink if you keep yourself hidden away during daylight?

'Well, think about it,' Amy said. 'Where would something that is bright pink and green hide? On something bright pink and green, of course. The elephant hawk moth's colour isn't for show; it's for camouflage. It roosts around rosebay willow herb, which is green and pink, and this is also the plant on which its caterpillars feed.'

Back at the car, Amy rummaged around the petrol-scented generator and pulled out two tubs, announcing, 'Now it's time for the beer and wine.'

A little early for me, I demurred. But as she opened the tubs it became clear this was not for us. Suddenly the odour of petrol was eclipsed by the smell of a brewery. 'We'll start with the beer,' she said. 'Beer before wine and you'll feel fine. Everyone has their own recipe; now let's grab some brushes and get painting.'

As we coated tree trunks and fence posts with this viscous brew, Amy explained that this was another effective way of luring in moths. It's known as 'sugaring'. So what's in her concoction? 'A very simple mix, this: a can of Belhaven Best, a bag of sugar and a tin of treacle. The vinous version just has a bottle of cheap red in place of the beer. This is last year's, in fact; maturing well, I reckon. Moths flock to it, feed on the sugar and might even get a little drunk on the beery good-ness; then they just wait for me to come and have a look. I hope they do that tonight, but I'm beginning to get worried that they won't appear.'

And with that we retreated to the car to chat about Amy's plans for a trip to South Korea, and wait for the light and the sugar to do their business. It was also a chance for me to investigate Amy's sudden loss of confidence in our night out.

'Didn't you feel it?' she said, a little incredulous. 'The weather forecast was perfect, overcast, mild; but look up, what do you see? A brilliant sky, inky black, poetically studded with millions of stars? Gorgeous, isn't it? Well, no it's not, it's a bloody disaster.'

For the first time I felt the pizzazz ebb out of Amy. She had been resolutely upbeat all day, but as the temperature plummeted, so did her spirits. 'Most moths don't fly when the temperature drops below four degrees centigrade,' she said.

After a couple of hours she relented, started the engine and put on the heater. But still, after another couple of hours we were cold, deeply cold, and decided that whatever we found now we would take as a night's work. Opening the car doors we found that we had been pleasantly insulated from the crisp night air. It was beautiful. Briefly, the peace was polluted by an aeroplane, but otherwise it was just us, the stars and a hint of frost on the tips of leaves.

First we visited the sugared sycamore, and were amazed by the life that had emerged to feast on the Belhaven Best mixture. Besides the woodlice, the snail and a harvestman, there were hundreds of tiny millipedes. So at least there was invertebrate life out and about at midnight. But as we passed along the line of posts and branches, there was nothing else, until, on the very last smear of beer and sugar, we found a caddis fly. This flying insect is different from the Lepidoptera, in that it has hairs on its wings rather than scales. It was not a moth. With sinking hearts we headed to the light trap. We were both beginning to shiver. We needed a little excitement, some adrenaline to knock the cold away. And as we crouched over the trap, Amy easing out each of the ten-or-so egg boxes, a little frisson trickled through me. Nothing, so far, but then we got the last box. We had both stopped breathing. There had to be something special waiting for us.

Nothing. An absolute abject failure of a mission to find the most common of animals. Dejectedly we packed up and wandered back from the inky blackness of this lonely field. There were, undoubtedly, thousands upon thousands of moths there, in among the vegetation, smirking.

Nearing the car, I suddenly noticed something moving beside the bushes to my right. 'A moth,' I called, and Amy rushed over.

'Ah, at last,' she said. 'Okay, it's a geometrid, one of the November moths. Given the temperature tonight, that seems appropriate, even though it's early September. I'm not going to take it home to identify it more thoroughly; that would be rather insensitive, since it's the only one that seems to have taken the trouble to get up tonight.'

Subdued, we headed back to Edinburgh. But Amy wasn't one to stay down for long, and soon she was regaling me with tales of her best nights of mothing, sitting outside a white-painted pub bathed in bright light, drinking beer and enjoying the spectacle as the moths swarmed to the lit wall. This is the sort of wildlife safari we can all take a part in, as long as there's a little warmth in the air.

8. DOLPHINS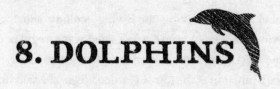

'Half a ton of sleek muscle'

After the disappointment with the moths– and it was a disappointment, how ridiculous is it to find just ONE moth when out hunting for them – I didn't have high hopes for my next quest, for a far more elusive beast. Taking advantage of being in Scotland, I had arranged to head further north. My instructions took me to Aberdeen and then to the bus station. Bus stations can bring out the worst in a community. I am not sure why, but they seem to receive less attention than any other mass transit terminus. There is never enough space, or it is empty and lonely and grey.

In such a dull environment it can be easy to spot something out of place. At first it was just the eyes. They had a twinkle of wildness and life in them. Maybe tiredness had beaten down my natural shyness, but I felt moved to say hello. And as I did, I noticed a logo on her blue fleece of a leaping dolphin and the letters CRRU.

She was one of us. Isabel Cordoba was heading up to the Cetacean Research and Rescue Unit, too. It was the shortest long bus ride I have had. We chatted and shared food and stories as we bumped along the B-roads between Banffshire and Banff, where we were met by the CRRU Land Rover for the final leg.

Isabel was researching the diving ecology and foraging behaviour of coastal minke whales and was a regular visitor to the CRRU. I had never been anywhere quite like it. As we neared a junction at the top of Gardenstown, she said that this was the last point where mobile reception was available, so I put in a farewell call to my family down south. And then we were confronted by the narrowest of lanes.

That might not sound so dramatic, but this lane went along the sea wall and was only one car wide. On our left was the row of houses that comprised the old village, on our right a three-metre drop to the banks of rotting seaweed. Izzy was greeted like a long lost sister by a stream of women who were waiting for us. My host and this animal's ambassador, Dr Kevin Robinson, had very nearly forgotten an important date. It was only when a member of the local Women's Institute had bumped into him in the shops and mentioned how excited they all were that he was coming for dinner, as well as giving a talk to the WI, that he realised he was double-booked. So he wasn't there when I arrived; but the five (female) volunteers were there, as were the four (female) staff, including my travelling companion. Being outnumbered can be quite fun.

Over the last fourteen years, Kevin has developed a self-sustaining organisation that allows him to do the work he is passionate about and share the experience. Self-sustaining might be an overstatement, though, given the tightrope he walks with his credit cards. It was clear that he was happy, had somewhere to live and enough food to eat; but it was a fine line. The way the CRRU operates is quite simple. Each summer around ten teams of people come up and spend twelve days working and learning. They get bed and board and pay enough so that the project can keep functioning.

I was not bringing money, but I did bring gifts of good whisky, chocolate and harissa.

By the time Kevin returned from the WI we were just getting into planning the next day's schedule. Inevitably, this sort of work involves being out at sea. I knew that. But somehow the reality of it had not sunk in. It's not that I'm *frightened* of the sea. Far from it, I adore being out on the water. But experience has shown me that this adoration lasts for about twenty-five minutes. Then terrible, grey-lipped seasickness descends. And I was also realising that the combination of beer, wine and whisky we had been enjoying was not going to help. I slipped into my sleeping bag feeling a mix of excitement and terror.

The next morning the people at the CRRU revealed the near-military logistics that are required to keep such an operation afloat. I was called in to get fitted for my survival suit. Not knowing what to wear, I had brought waterproof trousers, Wellington boots and a wax jacket. There was some barely stifled giggling, and it was gently pointed out to me that I might as well be wearing boots filled with concrete. When I fell in they would fill with water and drag me down.

First there was the woolly bear. Everyone should have a woolly bear. My kids have them and I have always been jealous of these snug, all-in-one fleeces. Then there was the talcum powder. Lots of it, spread around the rubbery seals of the bright-orange survival suit. This was partly to keep it in good condition, but mainly, I learned, to assist the re-birthing process, as my exceptionally large head went into conflict with the collar that would keep the water out.

More relief was on hand from Izzy. She sidled up and offered me a small pink pill. 'This is good stuff,' she said, with her heavy Spanish accent. I was not asking too many questions by this stage. The prospect of eight hours of sea-sick torment had made me decide to take whatever I could get my hands on.

Everyone was checked over and kit assembled – cameras, waterproof notebooks and lunch. On the drive to the harbour Kevin began to explain the work. Over the last fourteen years he has learned to identify individual bottlenose dolphins by the shapes of, and the markings on, their dorsal fins. So how can we volunteers help?

'This is where the cameras come in. If we can photograph all the animals we see, then we'll be able to identify them later. It also means I don't have to be everywhere at once. And a team of people like this means so many more eyes. Everyone has a job, as you'll see. Someone's driving, then we have at least one note-taker, plus photographers and spotters. Though sometimes we can be out there all day and see absolutely nothing.'

Kevin had begun to scowl at the state of the water. There was obviously a conflict: the volunteers want to get out on the water, but if the conditions are too choppy it becomes very difficult to spot dolphins and pretty much impossible to photograph them. With his vast experience, Kevin knew that there was a chance today would be a wash-out, but his is a strange position, straddling the role of scientist and that of Butlin's Redcoat.

The harbour is along the coast, west of Banff. There was a compulsory toilet-stop and then we started to load up. Two boats, one with five crew and ours with six. Kevin bought the rigged inflatable boats from an oil company. Known as Zodiacs, they're tough vessels, and the one I was in, *Ketos*, had a new engine. It started up with the sound of a sports car, a real throaty chug, as we navigated the collection of sailing boats in the harbour. But that was nothing to the moment we hit the open water. With the wind behind us, Kevin pushed forward the throttle and the noise was exhilarating, as was the rush of wind and the spray of water.

And then it dawned on me. How were we going to do this? The North Sea is very big, even the patch of the outer Moray Firth we were concentrating on dwarfed the boat. And why did we have what appeared to be harpoons?

'We know that these animals tend to follow a reef along the coastline, and typically spread out to feed in sandy bays,' said Kevin, when the engine-noise had lessened. 'We get help from other animals, too. In fact, you see all those gannets at about eleven o'clock?' I did see them. And what a sight. There were hundreds diving, and as we got closer we could see that some were barely getting airborne before plummeting down again. There seemed to be two species, brown ones and white ones. 'Chocolate gannets are this year's. They're full-sized, just brown. Anyway, they've found a school of fish, probably mackerel. Which is also what the dolphins will be feeding on, so we always investigate these feeding frenzies. Look, you can see on this.' He pointed towards the radar gadget by the boat's controls. 'This device is a GPS, as well as a depth reader, and a fish-finder – you see all those dots on the screen? They're fish, and they're at around three metres depth.'

Everyone was now busy scanning the surface of the water. If there were dolphins, they would need to surface to breathe. And while they can remain submerged for twenty minutes, while moving and feeding they tend to take a breath every thirty seconds or so. Kevin cut the engine and we began to do something that I was dreading. We wallowed. Swell, I love a good swell, as it builds waves to catch for surfing. But sitting as still as the swell will allow, bobbing up and down, this is the beginning of my end and we had only been out for fifteen minutes. I forced myself to stare at the tilting shoreline.

Nothing, other than gannets; and, for the first time, I had an understanding of how their name became a synonym for a greedy person. There is something quite relentless about the way these birds plummet. Amazing, too, the way their long wings folded back, so that the bird was arrow-like at the moment it entered the water.

Time to move on and the two boats separated. We, in *Ketos*, were staying closer to the shore, while *Orca II* was heading further out to sea. We were unlikely to use the 'harpoon'. The long pole was a mechanism for attaching a tracking device to the back of a minke whale. It does not happen often. Not only do you have to be lucky enough to spot one, but then you have to get close enough to poke it with the stick, attaching the suckers of the telemetric recorder to its back. After a day or so, the recorder detaches and the boats are scrambled to find it again and collect all the lovely data.

Moving on west along the coast we passed few other boats. But the ones we did pass gave a friendly wave, and it reminded me of driving in the countryside. There was an expectation of cordiality. But these interactions were more than just a social nicety. They were the result of vast amounts of work on the part of Kevin and his team. 'To begin with there was a worry among the fisher-folk that we might be a threat. Conservationists haven't always got the best reputation out here. But we have talked to them. We've tried to teach them what we know about the cetaceans, and at the same time learn from them what they know. No one knows more about what goes on under these waters than the people who are out twice every day, dropping and pulling creels.'

And the key message that Kevin had been pushing was that there can only be whales, dolphins and porpoises if there are fish. That is not to say that Kevin's dolphins are universally adored. 'They might be an important indicator of the productivity of the Moray Firth, but they are not always popular, especially when they hang around the mouth of the River Spey. People are paying up to £500 a day to fish for the salmon that the dolphins get for free.'

This understanding of the complexities of the relationships between wildlife and humanity is at the heart of what the CRRU does. It aims to work in the real world, identifying the key stakeholders and then trying to navigate a path through the minefields of competing interests and out into the safe water of common sense. Oil exploration, military exercises, over-fishing and tourism all have potentially devastating impacts on dolphins.

Oil rigs are a particularly obvious threat. Oil spills do not agree with wildlife. And as part of the bigger picture, they are also fuelling the change in the climate that will lead to rapid and uncontrollable shifts in the quality of the marine environment. But a less obvious problem is noise. The world of the dolphin is a world of noise. The use of seismic surveys, for example, at best drive cetaceans from optimal areas, and at worst cause death through disorientation and stranding. The noise made by the construction of off-shore wind farms is also a potential hazard.

Perhaps the most persistent noise problem is from the use of sonar by the military. For a long time the relationship between this and whale damage was hotly contested. But now the evidence is clear. So again, marine scientists like Kevin are in a position where they can influence what is done, to some extent, at least. Kevin can talk to the inshore fishermen, but influencing the multinationals and their miles of nets floating like walls of death in the oceans is another matter altogether. By-catch, the accidental catching of animals in nets, is thought to kill millions of cetaceans, the whales, dolphins and porpoises, worldwide.

On the boat the weather was definitely turning. What had been just a swell was now flecked with white. The wind was picking up and tipping the tops of the waves into breaking. The return journey was going to be a little more demanding than the outward one.

I got myself up to the viewing platform on the prow of the boat. The elevation increases the distance at which a dolphin can be seen. It dawned on me that I'd been out for nearly two hours and I was still enjoying myself.

The radio crackled into life. It was *Orca II*. They thought they'd seen a fin. They'd swung up ahead of us to a rocky headland. Kevin called his crew to attention. Cameras came out, waterproof note-boards were readied, and I sensed that it was a good time to hold on a little tighter. The rope around the side of the boat became my lifeline as the throttle was opened.

We saw the other boat and at that moment something spectacular happened. Half a ton of sleek muscle flew out of the water. We were too far away to photograph it. The other boat, we found, had no idea it was about to happen, either. I sensed a real quivering of excitement. I'd just had my first sighting of one of the iconic wildlife images, a breaching dolphin.

But Kevin was not satisfied; we got close to the other boat and found that they had not managed to identify the group that had been around them and was now around us. Both boats were bobbing and eleven pairs of eyes were willing the ocean to provide a hint of where our dolphins were.

A call went up – nine o'clock! – and there, to our left, were three dorsal fins just retreating beneath the choppy grey. I was concentrating hard, as if the more intensely I looked, the greater the chance of a dolphin appearing in my line of sight. I found that I was quite lost in the moment. This was a rare place to be.

The pod, or school, of bottlenose dolphins is very flexible. There is not a family unit; rather there is an ever-shifting pattern of groupings. Sometimes dozens may congregate at a rich food source, sometimes there may just be a mother with her youngster, or some roving gangs of males.

Rather too soon the school was gone. We had lost track. Each angled wave was a potential dorsal fin and it was clear that the deteriorating conditions were not helping at all. So we headed for home. It had not occurred to me that sitting on a boat would be quite so exhausting.

But heading into the wind meant we were climbing up the approaching front of each and every wave and then skiing down the far side. Spray was soaking us and we were consuming mouthfuls of the North Sea, when Kevin spotted a gannet storm close to one of the rocky headlands. He turned a little to the right and started gunning the engine towards the flock. All the time he was also keeping an eye on his magical sonar device. But he was not checking for rocks, he was checking for fish.

He switched off the engine and we started to bob towards the birds. 'Let's get the rods ready,' he said, smiling. And after a flurry of activity, two fishing rods appeared. I had been expecting some scientific device. No bait, just shiny metal tags with hooks. 'They're at three metres, just drop the hook in and you'll catch one.' So one of the volunteers dropped the hook in and within ten seconds had a tug on her line and reeled in a magnificent, sleek, iridescent torpedo. Ted Hughes sings a song to the mackerel, praising its 'plenty'. And each cast of the line brought another fish over the side, until Kevin had dispatched one for every person, apart from me. Over twenty-five years ago I stopped eating meat. For a good chunk of that time I was vegan as well; no eggs or dairy. The reason was always simple: the way animals are treated in the industrial production of meat is obscene. Any meat eater who refuses to acknowledge their complicity in the cruelty meted out on their behalf is contemptible.

Yet at no point in the last twenty-five years have I stopped craving meat. And here was meat, fresh, instantly dispatched, sustainably harvested and, from what I'd heard, delicious. This was the most ethically sourced meat I could imagine.

But still, when they ate their catch later that night, I chose the alternative – halloumi.

I slumped a little after that. Tired. Maybe also beginning to be affected by the bobbing of the fishing detour. But before nausea could rear up, Kevin asked if I wanted a go at driving the boat. Always up for some fun, I shuffled across to the inflated sausage of a seat for my lesson. Push lever forward, go faster; oh, and steer. Easy.

Now I like to think of myself as a cautious driver. Speed limits are respected, and I am very conscious of the fact I am driving a machine that can cause death to other road users. But the moment I opened up the throttle, heard the gurgle of the outboard erupt into a roar, felt the spray whip across my face as I accelerated into the weather, I morphed into a Jeremy Clarkson-esque petrol-head. The sheer exhilaration of the acceleration and speed, the closeness to nature and the wildness of the experience, conspired to super-charge me with profound glee.

We took to the air, again, as I failed to judge the oncoming wave and we crunched into the surprisingly hard water on the other side. Grinning like a mad fool, I pointed out to Kevin that I had never done anything like this before. But I noticed that not everyone else on the boat thought this was necessarily a thing to celebrate. And while I was at little risk of colliding with another sea-user, the chances of jettisoning my fellow crewmates were increasing all the time.

There is an art to travelling into the weather. And when it worked, it was quite a wonderful thing to almost surf down the back of the oncoming waves. But it required concentration;

all the more credit to those who regularly risk life and limb in their Zodiacs.

I had first come across the Zodiac as a tool of environmental protest. Greenpeace use them with great skill, dodging in between harpoons and whales, or evading the police. My wife is a video journalist and was out on one of their boats when the police tried to get them to stop. The delight when they realised that their boat was just so much more powerful, and that when told to stop, well, they did not need to, was great.

American novelist Neal Stephenson wrote most effectively about the combination of boat and protest in his 1988 eco-thriller, *Zodiac*. Who can resist such an unashamed protest group as the one he conjured up, called GEE, the Group of Environmental Extremists.

We were not fighting demon polluters today, we were fighting growing waves; and after a while I handed the responsibility to those with more experience. I reckoned we would get home quicker that way.

As I stepped off *Ketos* and onto the dock, a whole new world opened up to me. It was like walking across a bouncy castle. And I was happy. So happy that I had seen three dolphins; so happy that I had discovered the joys of speed, but above all, delighted by the effects of the little pill Izzy had given me. I had hardly felt nauseous at all.

The weather the next day was miserable. But the team was eager and everyone pushed to get up and out. Lunch was packed, suits readied; but Kevin came down from checking the weather forecast again and said 'Stand down'. The sense of disappointment was palpable. But the poor weather didn't leave the volunteers bored. Far from it, a great chunk of their time is spent processing data. So while they busied themselves at the array of computers, I headed out for a walk along the coast to Crovie.

This small village was perched on a cliff ledge so narrow that there was room for just a line of cottages and a footpath. No cars ever trespassed into the peace of this place, which was largely unaltered since massive storms in 1953 drove the residents away. A few people moved back and other properties were let in the summer. It was a dreamy setting, but one that must become a little exciting when the waves build and threaten destruction. Which is why most of the buildings were blind to the sea, sturdy and squat, with their gable ends facing the water and small windows shuttered against the autumn storms.

As I sauntered back I kept an eye out to sea. Among the rock pools an improbably billed curlew probed the sand beside an oyster catcher stabbing with its bright orange dagger-bill. The curlew took flight with its call that is so evocative of wildness, a weeping whistle that rose excitably above the seaweed.

There were patches of plummeting gannets, and then, suddenly, a different sort of movement in the water. At first it looked like a wave heading in the wrong direction, but then it was clear. A dolphin. In fact, three adults and a juvenile. What was more, I was standing on the one bit of the coastline with phone reception, and the dolphins were moving towards Gardenstown.

By the time I returned to the CRRU office the drama had long-since past. But on receiving my call, the office had emptied, as telescopes, binoculars and cameras were grabbed and the team piled onto the seawall to catch a glimpse. Even Kevin had been excited. Fourteen years working out here and he still had a puppy-like enthusiasm for what he was doing. And this infected the volunteers.

They were all crowded round computers, concentrating on photographs. The best images of each encounter were cropped and improved, sharpened and straightened. Then they had to work out who each fin belonged to. Some were easier than others; some had distinctive features such as a gouge or a scar. Other distinguishing features were more subtle. Hours can be spent poring over the images, trying to match them up. To most of the volunteers' frustration, Kevin only has to glance at a fin to know who it is. But the time it would take for him to work through the images on his own would mean he would only get out on the boat once or twice a week. There have also been some advances in technology, a bit like facial-recognition software, that can automate the effort.

Part of the deal is that the volunteers get education as well as entertainment, and after lunch Kevin began a lecture on the importance of photo identification in the study of cetaceans around the world. The fact that many species have dorsal fins that carry 'signatures' allows data to be gathered about individuals. And while this data in itself is fascinating, he explained that 'we no longer have the luxury of studying these animals just to learn more about them. Now the protection of these species is dependent on our learning more about them.'

All the CRRU's studies are driven by conservation initiatives. The day before I arrived, the team had popped a hydrophone into the water. Everyone was amazed by the onslaught of whistles, clicks and trills of the dolphin's underwater world. But Kevin was, reluctantly, unimpressed. 'It's amazing what we hear. But while the sounds of dolphins are undoubtedly interesting, how does documenting the acoustic repertoire of these animals actually help us in their conservation?'

Photo ID can generate data that directly influences policy. This is an area that Kevin is very keen to develop. So he is bringing stakeholders together and sharing his knowledge. The team at CRRU holds monthly lectures throughout the summer in Banff. And they are not just focused on communicating the results of the research to the audience – an audience that includes members of the public as well as members of the military and the oil and fishing industries. The lectures are also an opportunity to help people understand the scientific process and language. 'For example, when I say that something is "significant" I mean something very different to what a politician or a journalist means when they use the word. For me, "significant" is a robust description of the statistical likelihood of an event. For others, it might just mean it's important.'

An increase in the understanding of what scientists like Kevin are talking about means that the various stakeholders are more likely to collectively make good decisions with practical consequences. 'Using a scientific approach with photo ID allows us to build a picture of when young are born, for example, and where they are likely to be at certain times of the year. This can then be used to influence where and when dredging activities are undertaken, or where an offshore wind farm is placed.'

But the practical applications of the work do not completely overshadow the sheer pleasure of discovery. For example, while talking about ID marks on the minke (the seven-metre-long baleen whale Izzy is studying), Kevin suddenly drifted off into the wonder of it all. At the beginning of the week this team of volunteers had seen a minke breaching, leaping high out of the water on a mirror-calm sea. Fine and beautiful, you might think, but Kevin had just told everyone that minke whales didn't breach. He is learning all the time, too; he'd also just learned that individual whales appear to display lateralisation, or 'handedness', in their preferential feeding behaviours.

After another spell of checking pictures, we all headed off for a mad walk to the gannet roost at Troup Head. Even though I have seen this sort of thing many times before, I am always amazed by the ability of these birds to even perch on some of tiny ledges on which they make their homes. How they manage to lay and incubate their eggs there is just a wonder.

Another wonder was Kevin's belief that, despite the horizontal wind, we would feel comfortable creeping inches from the sheer cliff edge. Fifty metres may not sound very high. Trust me, it is, especially when the wind is threatening to tip even the stockier members of the troop onto their behinds. I decided to leave nothing to chance, and scootched along on my already-sodden trousers. Somehow I managed to get myself to the edge and peer through the telescope at one of the chocolate gannets sitting calmly a few metres away. I had assumed that little would be gained from using a telescope at such close range, but the transformation was incredible. I could even see the individual downy feathers above the bird's beak, which gave it a softer appearance than the white-headed adults who glam up with a bit of blue eyeliner. Another great wildlife encounter, and I had only been with the team for two days. I decided to return the favour (or take advantage of a captive audience), and after dinner I gave an illustrated talk about hedgehogs to one of the most unruly crowds I have ever encountered (and I've done a lot of WI groups in my time).

The next day was pack-up-and-go time for the volunteers. Their twelve days were over and some were clearly emotional about leaving. The impact that this sort of experience can have is impressive. 'Sometimes we won't hear from a volunteer for several months after they leave us, and then we get an email out of the blue saying that they've quit their job and gone back to university to study marine biology.' Kevin is obviously delighted with that sort of result. 'We have had a number of corporate groups over the years, where companies would send folk as a reward for their performance. One particular company would send people from their green teams. Each year we would ask about the ones who came on the previous trip. And often would get the answer that they had left. In fact, the head accountant of one of the companies quit and went off to Yorkshire to set up an organic farm, leaving everybody stunned. I wouldn't like to take the credit, because I wouldn't like to take the blame, but being up here is certainly a reflective and introspective experience for many of the souls that join us.'

Is this the way to undermine the system of global capitalism that seems obligated to damage life on earth? Expose company directors to the wonder of it all then send them back into the machine to work magical sabotage? I asked Kevin which bit of the experience really made the difference. 'Well, I have thought about that a lot. The place, here in Gardenstown, is magical. And the sea – being out on the water in those boats is thrilling. But there's also something quite meditative about all that time staring at the sea.'

And this got me thinking. There is a wonderful moment in the opening of Herman Melville's great novel, *Moby Dick*, where Ishmael describes 'the crowds of water-gazers . . . thousands upon thousands of mortal men fixed in ocean reveries . . . surely all this is not without meaning'.

A Buddhist friend, Rowan Tilley, had told me about 'mindfulness' – the art of 'bringing one's complete attention to the present experience on a moment-to-moment basis'. Back in Oxford, I described a day out on the boat to her. 'This sounds very like "access concentration",' she said. 'And the deeper you take this, the more likely you are to end up moving through "rapture" and "bliss" and then climaxing in "peace".'

Maybe CRRU ought to be re-targeting their marketing? It sounds like they are offering a shortcut to enlightenment in the guise of ecological research. Kevin is more pragmatic, though. 'Actually I think the decisive factor is the enthusiasm we all have for the work we do.' And he has a good point; seeing Kevin and his team beaming with delight and loving every moment must be very seductive for visitors.

After the last of the volunteers had gone and we had cleaned the guesthouse, Kevin looked at his watch, looked at the sky and the waves, and then, making a little show of indecision, said, 'How about one last run in the boat?' It was getting to the end of the season and the team were soon to disband. We were out on *Ketos* within an hour. On the way I had nudged up to Izzy and asked if she had any of her magic pills left. Equipped with pharmaceuticals, I was, for the first time, excited about the prospect of spending hours on a boat.

It was not quite a mirror sea, but the swell was gentle and the wind slight. Accelerating out of the harbour still gave me a thrill, and we headed back west along the coast, again following the reef. Despite the lack of volunteers, this was not a 'jolly'. We were still here to work, and twelve eyes scanned the horizon as, inadvertently, we drifted towards rapture. The hour before the shout went up passed quickly.

There were more dolphins this time, and we travelled with the seven adults and one youngster for twenty minutes before allowing them to continue their passage in peace. This was my first close encounter with dolphins and it was quickly clear why people rave about them. There is something mysteriously majestic about how they command their environment. Occasionally one would leap and this would be accompanied by the rapid-fire shutter clicks of Kevin's camera. Hadn't he got enough beautiful photographs? I asked. 'I'm trying to get a clear view of the animals' genital slits,' he explained. Was this appropriate conversation for polite company? 'We don't know the gender of our animals, even though we may have seen them many times and know them as individuals. Another way we can do it is when you see one with a calf, they are obviously a female, like here,' he said, pointing to the young-ster, which was clearly smaller, and quite pale in comparison to the dark grey adults. 'This is the first time we've seen this little one, a newborn, less than a month old, I'd guess. Now, we already know that Sooty, its mother, is a female because we've seen her with other young calves over the years – this is at least her fourth calf. We first photographed Sooty in July 1997.'

Long-lived animals such as whales and dolphins require long-term and ongoing studies. This is the mantra of the CRRU. They're not after a quick thrill or a quick fix. On the journey back to port I asked Kevin whether he'd ever swum with the dolphins. This seems to be an experience that many people would like to have. There are many stories of miracle cures as a result, as well as tales of dolphins rescuing swimmers from sharks. Bottlenose dolphins seem like entirely benign entities.

I can think of no other creature that has quite so attracted the 'age of Aquarius' as this one. Perhaps it is the constant, all-knowing smile? But Kevin, despite his passion for these animals, is not moved by this more esoteric interest.

'When people ask me this I tend to ask them if they'd go running with lions while out on safari. It's a similar thing. These are the top predators in these waters. They're very big. They can be killers.' But what about all the wonderful swimming experiences people have? I'd met a man who used regularly to swim with a dolphin off the northeast coast. 'There have been some detailed studies of these sorts of encounters, with what we call "solitary, sociable cetaceans",' said Kevin. 'What usually happens is that a single dolphin arrives in a new area. Being deeply sociable and inquisitive it is naturally drawn to boats, but it is generally people who initiate the first real contact. The press picks up on it, and suddenly there are hordes of visitors; the more boats in the water, the greater chance of injury and death for the dolphin.'

What is also not understood by many of those wanting to leap into the water with the best part of a tonne of muscular carnivore is that the dolphin's motivations are probably quite different from our own. It is unlikely that the dolphin is channelling a life force that will heal you and the world of all ills. In fact it's possible that behaviour that we identify as play might in fact be aggressive or sexual. Neither of those scenarios is going to end well for the smaller party.

As we were chatting there was another movement in the water. I caught a glimpse but before I could shout 'Dolphin!' Izzy had shouted 'Porpoise!' I leaped to my feet and a few seconds later the fin of the animal – much smaller than a dolphin – re-appeared for a brief moment, before rolling back down. I kept straining to see more.

The harbour porpoise had originally been the animal I wanted to see on this trip, but Kevin had said that they were not so easy to spot, and typically avoided close boat contact. I had wanted to focus on them because they are not as obviously charismatic as the dolphin. Everyone loves the dolphin; I thought it far too easy a subject. But the reason so little is known about the porpoise is simply that researchers get to see them so rarely, and then only for a short time. And this is despite there being vastly more porpoises in the North Sea than dolphins.

Porpoises are certainly not as 'cute' as dolphins. They don't smile as they go about their business, and their name comes from a conflation of the Latin for pig and fish. So not as appealing. But they are the first cetacean I saw in the wild.

When the water was calm off the island of North Ronaldsay – the most northerly of the Orkneys – and I had some time to myself between nights of hedgehog counting, I would head for the rocky shore to the southwest and sit. This was before I had read Melville's account of water-gazers, but that is what I was doing, letting the scene wash through me. On one of these excursions I spotted what I thought was a dolphin and headed back to the bird observatory in which I was staying to share the news. And that was where I learned that I had glimpsed my first porpoise. Other more experienced eyes had also been looking out to sea.

'You know, some of these gorgeous smiling dolphins are murderers,' said Kevin, seemingly out of the blue. 'They kill porpoises.'

Well, they are top predators, I thought. How did killing smaller creatures make them murderers? But Kevin explained. Over the years, they had been finding the bodies of porpoises washed up on the shore. As part of the CRRU's work, they recovered the carcasses of stranded cetaceans for necropsy. The body fat of these animals accumulates the persistent pollutants that get concentrated at the top of the food chain. Building up a picture of this and other threats that cetaceans face allows the CRRU to influence policy-makers who can push for changes in the way the waters of the Firth are used.

These autopsies revealed that a surprising number of the porpoises had been killed by dolphins. But why and precisely how still remained a mystery. Then tourists on a whale-watching trip videoed two dolphins killing and then seeming to play with the body of a harbour porpoise.

'So it is deliberate,' I asked. 'They are doing it on porpoise?'
It must be really difficult for Kevin to hear the same jokes
so many times. But it had formed in my head and exited my
mouth before any of the editing function in my brain had
kicked in. I should like to thank Kevin for not pushing me
overboard.

The two dolphins had worked together, one herding the
porpoise and the other striking it with its long beak. But why?

'A few years later, after we had seen the behaviour ourselves,
we saw something different that, in a quite unpleasant way,
gave us the answer.'

A female bottlenose dolphin with her youngster was
targeted by a pair of males of the same species. They herded
the mother and infant away from the body of the pod and
then concentrated on splitting them up. At which point
a prolonged and directed attack began on the youngster.
This was all done in the same manner as the attacks on the
porpoises. Kevin was clear about what this revealed. 'We now
believe the bottlenoses are using the porpoises for "target
practice". Baby dolphins are approximately the same size as
porpoises.'

But this only generates another question. Again, why?

'I mentioned lions earlier, and the parallels are greater than
their both being a top predator,' Kevin explained. 'A female
dolphin with a new calf will not be receptive for mating for
two to three years while raising her youngster. But if her calf
dies, she'll come into season within a few weeks. This is just
the same behaviour exhibited by a male lion entering a new
pride. He wants to ensure his genes get passed on and will go
to extreme lengths to make that happen. So he kills the current

crop of cubs, thus encouraging their mother to become fertile far sooner than she would otherwise.'

This was bubble-bursting information. These animals are undeniably one of the most charismatic species in the UK. But like charismatic people they are complicated. Perhaps that is why I am so attracted to the simpler life of the humble hedgehog. And in that moment my mind is made up. No dolphin for my leg.

Back on land, exhausted and exhilarated, with a salty face, sunburned nose, matted hair and a glass of fine whisky, I collapsed onto the sofa next to Kevin. 'You know, we need a name for that youngster,' he said. 'Would you like the privilege?'

What a responsibility. Immediately there was a problem. If I used either one of my children's names, Tristan or Matilda, the other would be furious with me. And then there was the question of gender. A non-gender-specific name would be good. But what? Then it struck me – a way of forming an attachment with this animal. My website is 'urchin.info' – because someone got to 'hedgehog.info' just before me, and urchin was the closest I could get. How about 'Urchin'?

So I am now the adoptive father of this remarkable dolphin, and I really hope to get a chance to go back and see how it is doing. I hope that it manages to avoid the many pitfalls that await it, in its 'dolphin-eat-dolphin' world. I hope that it will thrill hundreds of people as it makes occasional appearances for its ever-loving public, helping them on their way towards enlightenment.

Because there is enlightenment to be gained from time

spent in the company of wildlife. There is a calmness that seeps into you, despite the obvious excitement, a state that really is, as my Buddhist friend described it, moving towards rapture. Oh, and the wonder drug? Biodramina.

9. OWLS

'A predator hunting on our doorstep'

I had high hopes for owls. My ambassador, my guide, was one of the most articulate and experienced owl experts in the country. As Conservation Officer of the Hawk and Owl Trust, Chris Sperring, MBE, is at the forefront of communicating about these amazing animals. But better than that, it was an obsession for him, one that started when he was a child. Better still, he was the guitarist in a rock band called Raptor.

I met him at his home in Portishead, just outside Bristol, where, after a quick cup of tea, he said, 'Right, let's go and find some little owls.' Clambering up into his 4x4 we bounced off towards what he described as 'Classic little-owl habitat, absolutely classic'.

There was no mucking about, no beating around the bush, as we sized each other up. Chris is a professional and was feeding me information from the outset.

To begin with, the rock band. I had been hoping to come to a gig or rehearsal, but Raptor had suffered a setback on their sure path to global domination when the drummer moved to Brazil. But Chris was not downhearted. 'I'm planning something special; recruiting some new musicians,' he said. 'It is the Somerset Wildlife Trust's fiftieth anniversary coming up, and I'm going to take the new band on a county-wide tour, raising money for the charity and raising awareness about the amazing wildlife we have here.'

He pulled the car off the road and jumped out. Grabbing recording gear, cameras and binoculars, I joined him as he strode purposefully towards a small humpback bridge, where he stopped and gazed expansively around and repeated, 'Classic little-owl habitat.'

I followed his admiring gaze. The view was a little unprepossessing. The bridge on which we stood once carried the road over a railway line, before Beeching's axe fell and these branch lines were closed in favour of the motorcar. Chris and I stayed close to the bridge wall as cars bombed along the B-road. Across the fields the M5 droned a dirge for the land it had destroyed. Despite the early hour the industrial landscape was already drenched in electric light. 'Unprepossessing' was generous.

'Stop looking at it like a human,' Chris said. 'Look at it like a little owl. Think about what the owl needs to survive. This pasture is given over to horses now, and although the fields are a bit bigger than they once were, there are remnants of hedgerow and a few fully mature oaks. The cracks and crevices of an old oak are the ideal places for a little owl to nest. And the horses produce something very special that we in the West Country call S-H-One-T.'

And the reason the outpourings from the horses are so valuable is that they attract little-owl food – beetles and other coprophiliac invertebrates. The little owl is, unsurprisingly, smaller than Britain's other owls – about half the size of a tawny owl, for example. And they are delightful. I remember, on a long bike ride through the Cheshire countryside, somehow managing to spot one perched in the crook of an aging oak. I stopped and entered a futile staring match. The owl's flat head accentuates its serious eyebrows, which make its yellow eyes all the more mesmerising.

'I've lived around here all of my life and there have always been little owls. Let's see if we can rustle one up for you.' And with that, Chris pursed his lips and let out a sharp curling whistle; then he paused and did it again, all the time concentrating on the fields in front of us. We waited, but nothing. So he repeated the whistle. He was hoping to elicit a response from an owl. 'There's a problem for wildlife around here,' Chris explained as we waited. 'If we were to jump into a time machine and go back just thirty years, you would be able to appreciate the changes that have occurred, and understand what is happening to the countryside and the wildlife. We're putting so much pressure on what is left. Over 6,000 houses have been built around Portishead since I was a kid, pushing the town right up to the docks, squeezing the buffer of wildlife-friendly ground down to nothing.'

Even the horses are far from harmless. They graze the grass so close to the ground that it's left looking like a football pitch. 'I used to look at everything solely from the perspective of the owls,' Chris continued. 'Now I also see the value of those thistle outcrops, for example. They're providing a winter seed-store for finches. Just looking at one species on its own is a waste of time; you have to understand the ecosystem, how everything is interconnected.'

Chris tried whistling again; nothing.

'Look, we'll go to the other patch where I know there are little owls,' he decided. And we headed back over the bridge, through some delightful carved gates into a nature reserve. As we walked, I asked Chris about the threats this smallest of owls faces. Is it a very specialised feeder?

'Actually, though their main food source is insects, they can take small mammals as well, which puts them in a stronger position than real specialists like the barn owl, for instance, which relentlessly feeds on the short-tailed vole,' he said. 'But they do suffer from being prey themselves. Feeding on the ground, as they do, makes them vulnerable to foxes, in particular.'

We had reached another gate and Chris climbed over it. As I joined him, he said, 'I know there's a bull among the cows in this field; just not sure where he is right now.' He sounded genuinely concerned. I asked him what his strategy was for dealing with enraged bulls. 'How fast can you run?' was his answer.

Finally we made it to the adjacent field. Beside the hedge we stopped and Chris pointed up to a large bird box. 'Barn owls nested in there this year; they had three young,' he said. There were white splashes of bird droppings all around, and a few feathers. 'The white, it's called "mute" staining or "whitewash". It's the excretions of the owl, and not that interesting, apart from telling you loud and clear where they've been roosting. What's fantastic to find are the pellets.'

The hunt for animal 'leavings' is getting to be a consistent part of this quest. But the pellets regurgitated by owls are special. Unlike most birds, owls do not have a crop where food gets broken down on its way to the stomach for digestion. The owl just swallows everything whole – mice, voles, shrews, down in one. Once inside, the digestive juices get to work on the meal, stripping the bones of goodness. Then the remains are cleverly wrapped in the fur before the owl coughs up the resulting pellet – for an ecologist to come along and dissect it. Because it is the pellets that tell us what the owls have been eating. There are entire guides to teasing out the differences between the lower jaws of pygmy and common shrews, for example. The one pellet-like thing I found turned out to be the remnants of fox.

The tree in which the barn-owl box was located was truly magnificent. The bark was richly wrinkled, like the skin of an old man. 'I love these ancient trees,' said Chris, joining me as I peered into the mini-canyons coursing along its surface. 'They have so many inhabitants; they are so much more than just a tree. They are an entire world for some animals.'

We were still on a quest for the little owl, and I wondered whether the two species got along. Would the presence of the barn owls reduce our chances? 'They live together quite happily. They're not competitors for food or space. Shall we give it another go?' he said, pursing his lips again and sending out another whistle; then pausing and whistling again.

A roe deer walked into view, cocking its head in our direction before calmly walking on. Then a blackbird kicked off with a hysterical babble. 'There's a little owl out there, that blackbird's given the game away.' But might it not have been his owl-whistling that frightened it? 'I think we're too far away. Last night I walked around here, and I'd seen three owls by the time I got this far. Let me try something different.' With that, Chris hooted again, making a much more stereotypically owl-like noise.

'What I was doing before was a contact call,' he explained. 'The noise I just made was a call they use in the spring when males are defending their territories.' He hooted again and I was left wondering whether there were little owls all over the field confused by the sudden arrival of a spring-fuelled male as autumn was beginning.

There was movement. A sparrowhawk zipped low across the hedge; this might well have been the cause of the black-bird's discomfort. As we readied to move on, there was another movement, this time on the ground. A magnificent fox nonchalantly gazed in our direction before trotting on. So rich and ruddy, very healthy-looking, unlike the rather mangy individuals that I see close to home. But still no owls.

'Come on, let me show you a special tree,' Chris said, unbowed by this failure to conjure up an owl.

Not far along the hedge-line was another old oak. 'This is the tree that started me on my career as an owl advocate,' Chris explained. 'Around 1972 – I think I would have been eleven – I was already obsessed with owls, and I had brought the chairman of the Somerset Wildlife Trust to this tree to show him the hole where the little owls were nesting. He liked my enthusiasm and wrote to John Sparks and invited him down. Sparks had written the book on owls that was my bible. He went on to be head of the Natural History Unit at the BBC in Bristol. And I can still remember the feeling of pride as I walked up to this tree with my hero to show him this nest.'

We turned to look back into the field and had just walked through when we caught sight of the fox again. It was swiftly joined by two more. 'Now, I reckon that is a mother with her two adolescents; let's see what happens,' and with that

Chris rolled up his left sleeve and pressed his lips to his arm. For a moment he looked like a school-girl practising kissing. Then a remarkable squeaking noise shot out and across the field. The result was immediate. The two smaller foxes both jumped up and looked towards us. Chris kissed his arm again and the two foxes started to run towards us in an obviously excited way. They were pronking like antelopes. One of them got to about ten metres away before it realised something was not quite right, and both of them continued to run in an arc back to their mum.

'What has just happened?' I had to ask.

'That,' said Chris, with a smile of accomplishment, 'was the sound of an injured rabbit.'

It was a magical encounter. I love this time of the evening, at the point of transition between night and day. When we are indoors this moment passes too quickly. As soon as it gets a little dark we flick a switch and on come the lights, obliterating the view of the outside world. But when you are out in it, in the moment of change, there is a great subtlety to the diminishing light. Adjustments and compensations in our senses come in to play as we hear more and see differently. Now is the time to stop staring at things you are interested in as the colour leaches from the day; we must start to shift from the cones of our eyes to their rods. When you look at something directly you are using the less sensitive but more detailed collection of light receptors in your eyes, called the cones. The more light-sensitive rods are not concentrated at that point of focus, so when the light goes we need to start to see the world out of the corner of our eyes.

And the magic of Chris's Dr Doolittle impersonations

filled me with envy. He had called the animals in to us. That is a powerful tool, and I just wish I could do the same with hedgehogs. It would make fieldwork so much easier.

But still no owls.

Walking on, I found we had done a full circle and were now back in the bull field. Chris took me back through the terror and on to the edge of a field with much longer grass. 'This is perfect barn-owl habitat,' he said, trying to sound confident.

Readying to record him making a squawk, hoot or whistle, I was amazed at the sound he produced. It was like the noise my radio-tracking gear makes, white-noise, a powerful *hsshh-shshshhhs*. He grinned at me. 'Barn owls are much less vocal than little owls, because they don't need to communicate in that way. They're very easy to see when they're quartering a field, even to us humans. And if *we* can see them, then for another owl, with eyes a hundred times as sensitive as ours, well, it will be like a beacon.'

The image of a barn owl flapping like a pale tea towel back and forth across a field near the Thames in Eynsham is one of the abiding memories of my honeymoon. We were sat on the decking of a small wooden house on a secret island, enjoying the last warm rays of the day and the first cold drinks of the evening. At first all I spotted was movement, but then we both saw it. As wonderful as lions stalking a zebra, here was a predator hunting on our doorstep. Who needs to go to Africa for a thrill of wildness?

The rise of reason and rational thought has put paid to many of the reactions such a vision might once have elicited – because owls are an animal of portent. The ancient Romans regarded the owl, with its nocturnal flight and eerie cries, to be a messenger of death. The Bible is riddled with owl-phobic writing, and it was frequently regarded as a witch's familiar. To be fair, there are also ancient allusions to the benevolence of the owl, and it is interesting how divergent different cultures' relationships with a species can be.

'Let's see if a dying vole will pull an owl into the open,' Chris said, readying his arm for more kissing. I pointed out that the squeaks sounded rather like the dying rabbit. He gave me a look that suggested I was in no place to challenge his authority on the noises of dying mammals. 'The only times I see absolutely nothing is when I'm with someone who's brought recording equipment,' he said pointedly.

'Okay, tawny owls are much better behaved,' he continued, 'but first we have to stop back at home.'

The short journey back was filled with a debate about the rights and wrongs of eagle owls. These magnificent birds, big enough to feed on hedgehogs, have recently been found to be breeding in Britain, and there is a bit of a fuss. Conservationists are worried that they might impact their programmes to protect and reintroduce other raptors, like the goshawk. The concern is largely due to their being either reintroductions or escapees that were originally captive. 'But there have always been eagle owls over here,' Chris argued. 'If you look at the bird-artist Thorburn's notes from Queen Victoria's day, he mentions them as an occasional addition to the game bags of Scottish estates. They've been drifting across the North Sea forever. It's just that now they're more established. And they're calling for a cull. It's utter madness.'

Chris's anger is obvious, and he's been working hard to

communicate the issue to MPs and ministers. 'The main problem is that none of them have a grasp of essential ecology. There was an eagle owl that took up residence in Bristol, an escapee, and he spent six months living around the university. Crowds would congregate to see him roosting during the day. And what did he eat? There was a pair of peregrines on the building next to him, but he didn't bother them. No, he fed on what was common: brown rats, pigeons and grey squirrels. He was doing us all a favour. Rather than trying to exterminate them, we should be encouraging them.'

Pulling up outside his house, we jumped out and he told me to go inside and make myself a cup of tea if I wanted, and headed off into the garden. I could hear a noise not dissimilar to the ones Chris had been making only a few minutes ago, so I followed it to find him in the back garden feeding day-old chicks to a barn owl. 'I forgot to feed her before we went out,' he said. 'She was hungry. This is my PR and Education Officer. She has been on more than 120 school visits and has probably done more good for owl conservation and general awareness about the importance of wildlife than any human being. Isn't she gorgeous?'

He had a point. Hand-reared and unable to cope with life in the wild, the owl now had a role of great importance as an ambassador for her species. Close to, it was amazing to see her huge eyes. Unlike ours, the owl's eyeballs are rooted in their sockets, so to see anywhere other than straight ahead they use their extremely flexible neck to move their whole head.

'I don't bother taking her to an RSPB talk, for example,' he continued. 'Those people are the converted. But when I go into inner-city schools, that's when she comes into her own. A lot of these kids have no chance of seeing anything like this in the wild, and for some of them, when they look into those wild, wild eyes, they are meeting something as wild as they are. It's a very special connection that can be made.'

Leaving the well-fed ambassador behind, we headed back out, ignoring the car and walking up into the woods that loomed above Chris's home. Stepping over the threshold, out of the suburban glow, was like having a velvet blanket thrown over our faces. We are not used to true darkness. Just as the drone of traffic is an ever-present imposition on our ears, so the eradication of night by artificial light imposes a great absence on our eyes: the absence of complete darkness. That is not a call for a reversion to pre-industrial life. If anything, a little bit of absolute darkness makes us appreciate the instantaneous light all the more.

This was definitely Chris's patch. These are the woods he walks in daily, or rather nightly. We had torches, but he was sure-footed along the paths, his feet carrying a memory of the obstacles and pitfalls, so we kept the torches switched off, for now. It takes time for our eyes to adapt to the dark, and while we can develop reasonable night vision within ten minutes, it takes a good half hour of darkness before our night sight is at its best.

Not far in and Chris stopped. 'I'll start with a male,' he said, and cupping his hands to his mouth made one part of the classic owl hoot: '*whowhoo*'. He paused, and repeated it. I was waiting for the sound that I expected to follow: '*twit*'. I could tell he was thinking the twit was me.

'The owl we're looking for is the most common owl in

Britain, the tawny,' he explained. 'And this is the owl that has taught every child how an owl sounds: "*twit-twoo*". But what most children don't know is that this is the noise of *two* owls – a male, who goes "*whowhoo*" and the female who gives a sharper screech, which is a bit like "*twit*".' Which Chris then impersonated – a great screeching whistle.

These birds are so vocal because they live in a much denser and darker habitat than the other British owls. They are also extremely territorial, and establish their territories using sound. Which is why what Chris was doing was such an effective way of learning who was living in the woods.

'This time of year, autumn, there are lots of young birds moving around,' Chris said. 'I'm playing at being two birds looking to settle down, so I announce myself to the wood and if there is a resident owl, it will tell me to bog off.'

And what happens if you don't 'bog-off'? I wondered.

'Well, the resident will move in closer, hoot or screech a little louder, maybe fly by and take a look. Most sensible owls will have left by then, but sometimes it comes down to a fight.'

A fight? Really?

'Oh, they can fight bitterly. I was told one story of a tawny owl that had been perched on someone's roof, calling constantly, when a second owl swooped in and knocked it down into the water butt. And it didn't stop there, the attacking owl then landed on it, holding it under the water until it drowned. Now, this is not common, but they do fight and I've seen it many times. The killing is often a little more subtle, though. In these woods I've seen two aggressive birds rise up into the air opposite each other, lock talons together and come crashing down to earth. If one gets a good grip, well, the other is as good as dead. If its wings are damaged in any way, it can't fly and becomes fox food. Even I have been attacked.'

He hooted again. 'You can't just leave it at that,' I complained.

'It's only happened twice. The first happened when I was very young. The second was rather embarrassing,' he said. 'I was out with a group from the local WI, watching owls at twilight. We were at the entrance to the wood and I had been making owl noises, and was then talking to the women with my back to the wood. They saw it come swooping in, but they didn't say a thing, and then it walloped me right between my shoulder blades. Interestingly, its talons weren't out, but it was a real thump.'

That owl must have gone back into the woods rather shocked at the stature of the interloper it had attacked, I suggested.

'It's not just owls that get confused,' Chris continued. 'I have it on good authority that someone with a passion for owls moved to a new area and tried his luck, mimicking owl calls as I have been tonight, and was delighted to get a response very quickly. Night after night he would go out and chat to the local owls. A few months later he had a house-warming party and invited the neighbours, and got talking about his passion for owls. It was only when he mentioned how lucky he had been to have found a resident near his new home that he discovered that one of his neighbours had been equally delighted when an owl had moved into the neighbourhood a few weeks back, and he had been chatting to it, too.'

I can see how good Chris is; every story has an opportunity

for another, and all the time, just a little bit of information about owls is imparted. And I was helpless, I needed to know more.

'New Year's morning, every year, people come and take a fairly organised walk through the woods,' Chris continued (and I was determined we were going to get back to work after this one). 'I was up a tree, just minding my own business, when I heard them stomping around below me, making so much noise that they were never going to see any wildlife, and neither was I for that matter. Anyway, the devil got into me and I thought I would have some fun with them, give them a treat. So I went 'cuck-oo' a few times. It was great to watch them, spinning around, talking, looking confused. Best of all, there was a letter in the local press about the earliest ever recorded cuckoo for Somerset.'

Another hoot and another screech from Chris, but to no avail. 'Never mind,' he said, 'let's head up to the quarry.' On the way, something occurred to me. If other people could be fooled by human owl-mimics, could Chris himself be fooled? Might it be skewing his research?

'I've been doing this for so long now that I don't think it would happen,' he said. 'In fact, I can recognise individual owls; some birds have distinct accents. It's easiest to recognise the ones with a defect, of course. Take great tits, for instance. They usually go "teacher, teacher, teacher", but some do it a bit wrong and go "teach, teach, teacher", for example. Now, that bird you will recognise next time you are in the wood. It's the same with owls, the better you get to know them, the more you can identify the differences in their calls.'

We suddenly emerged into space, as if we had left an

oppressively cluttered room via the French windows. The stars above sparkled in a particularly jewel-like manner. This was partly because our eyes had grown accustomed to the dark by now, but also thanks to the wall of the quarry fifty metres ahead of us, which blocked out the invidious light pollution. The air felt so very different here. And when Chris started to call in the owls, the sound bounced back from the quarry wall, giving it more life than it had had under the blanket of wood.

Again, he was met with a great big nothing. A pause, a repeat, nothing. Was there a moment? I asked. A particular event, other than his formative encounter with John Sparks, that had settled him onto this course?

'There have been many moments that have brought me to where I am, but there was one in particular that had a very big impact,' he said. 'From birth I was a naturalist. I was always interested in nature; in exploring, delving, climbing. I was eight years old and had learned how to make owl-like hoots. I had climbed up a tree near Portishead and was sat there, hooting pretty much to myself, when a tawny owl flew down to take a look, and settled on the same branch. I could have reached out and touched it. That was a moment of truth, when I knew I had found what I wanted to do.' Suddenly he stopped. 'Did you hear that? It was a female.'

I'd missed it; too busy concentrating on what Chris had been saying. He seemed relieved: there was life out there. Silence. I think we were both holding our breath. All I could hear was a cricket stridulating. Then a distinct but distant screech, and we could both breathe a little easier. Chris screeched in return, a much harsher sound, and I jumped.

'Just trying to wind her up a little,' he said with a smile. 'Let's go and see if we can get any closer.'

The wind was beginning to pick up, but above the rattle and sigh of branches we both caught a hoot. A male. 'Now, that means there is probably a pair here with an established territory. They're getting ready for winter and will work together to keep this patch theirs.'

Returning to the dark embrace of trees, we caught another hoot, this time coming from our right. 'We'll wait here,' Chris suggested. Then a screech. 'I think we're between them. I just don't know whether they've already come and taken a look at us and gone back up into the canopy. You see that patch of light,' he whispered, nudging my gaze upwards. There was a piece of sky, the fingers of the bare trees tracing a silhouetted filigree around its edge, like the fringe of a doily. 'That is where we will see an owl.'

The wind was lulling and the woods were dark and calm. No more chatter now; this was serious. Around us the trees creaked, as if flexing themselves into more comfortable positions. We waited. My mind filled with the sounds and smells and sights of the wood. But there was no more hooting or screeching, and the pale patch of sky revealed nothing.

Possessed of finer spider-senses than me, Chris tapped my foot with his, alerting me to something. I looked around, straining all my senses. Nothing. And then something extraordinary happened, or almost happened. I do not know whether the wing touched my cheek, or whether it was the breath of the wing. I felt the most delicate piece of moss-green silk graze the very edge of my sensible world.

My body did not move but my mind reeled, as the import

of that brush with the wild sank in. I struggled to find words; I needed to let Chris know what had just happened. Here was the wild coming to look at me. No longer the observer, I had become the observed.

'Did you feel that?' Chris asked. I nodded, then – realising the futility of that gesture in the darkness – managed to say yes. 'That was wild, wasn't it? All that time looking for them and then one comes to look at *us*. We didn't even see it, but it saw us alright.' And it was clear that even this seasoned owl-man was as exhilarated as I was by that interaction.

As we retreated from their territory and headed back to his home, Chris started to talk about some of his other work. 'I'm doing this series of lectures at the moment called "The Real Wild". I'm trying to get people to reconnect with the wild. We're missing connectivity. That is at the heart of ecology, the interconnectedness of life. And if we don't "get it", then we don't have a hope in hell of solving the mess we're in. Anyway, talking about the wild to these people, I found that I was thinking more and more about my son, Markie. He's the only one of my kids who will join me on these expeditions into the woods. That's partly because he can't argue against it – he's heavily autistic and doesn't communicate much. And I realised that he's been teaching me so much about the wild, because *he* is wild. He's not been "domesticated"; he reacts to what he senses in a purely instinctive way. If he wants to find out what something tastes like, he puts it in his mouth – whether it's berry, a leaf or a handful of mud. But he's never put anything dangerous in his mouth; he's never eaten poisonous berries or fungi, even though there's plenty of them around. He just seems to know. He's my wild animal. He helps me connect to the wild as much as the owls do.'

Back at my car, having met Markie and said my goodbyes, I was left feeling strangely moved by the experience. The quest for the owls had initially been just a matter of fun and education, but the evening had rapidly developed into something far more profound. I drove down to Gaia House, a Buddhist retreat in Devon where a friend was working, for a night of restorative calm. And as I lay in my narrow bed, in the former convent cell, the window (wide open) drew my slipping consciousness out into the green, red and brown of the autumn night and my last thoughts were of a conversation between a couple of tawny owls.

10. ROBINS

*'The liquid warble of a robin can bring
a little brightness to our hearts'*

Despite the profound delight of my owl encounter, birds have never had as much impact on me as mammals. I enjoy watching birds, but there is not the same visceral excitement as I experience when I see a wild mammal. But I had been looking forward to the day with my robin man a great deal. Dr Andrew Lack has many strings to his bow. While he is an academic and a well-respected and energetic lecturer, he is also an author, co-writing, for example, the Collins New Naturalist monograph on pollination.

We had last met on a beautiful, early summer's evening in the graveyard of Marcham's All Saints Church. This was the Oxfordshire countryside at its best; a village bathed in golden light and an audience packed inside the church, gently murmuring in anticipation. I wished Andrew good luck and went to reclaim my seat in the front row.

As leader of the Isis Chamber Orchestra, Dr Lack was already revealing a diversity of talent that would make even the most saintly a tad green. But there was more. Following on from the orchestra's performance of Ernest Moeran's 'Sinfonietta' there was a brief hiatus as chairs were moved.

Andrew Lack reappeared, to a wave of anticipatory applause. Holding his beloved violin with the firm tenderness usually reserved for infants, he smiled nervously. The conductor calmed the air with his wand and Andrew began. There was no escaping his lightness of heart as, with the first few notes, Vaughan Williams's Lark took flight.

The bliss of that evening was all but forgotten as I cycled towards his Oxford home. The rain was horizontal and bordering on sleet. Bedraggled and pretty miserable, I arrived at Andrew's home. I stripped off as much as decency permitted and the heating was turned up. Andrew suggested we settle down for lunch before getting into the thick of the interview. As we enjoyed leek-and-potato soup with bread from the farmers' market we chatted about music as much as wildlife. I love a passion, but even better is two passions. And while the details of ours differed, the essence was the same.

Eager to compete with his Lark, I told him about the delightful song I had learned, written in 1926 by Harry Woods, 'When the Red, Red Robin (Comes Bob, Bob, Bobbin' Along)'. It's a great song for kids, and I broke into a short rendition over the soup: 'Wake up, wake up you sleepy head; Get up, get up, get out of bed; Cheer up, cheer up the sun is red; Live, love, laugh and be happy . . .' I suggested that Andrew's orchestra ought to try this at its next performance.

He looked up and said in a gently concerned voice, 'You do know that this song isn't about our robin, don't you? It's about the American robin, a very different sort of bird: a flycatcher with a red breast. There's also a group of flycatchers in Australia called robins. We were so fond of them that we took the name with us as we colonised the world, and gave it to any bird that resembled our British species.'

There were also many attempts to introduce the British robin to the colonists' new homes. Like the hedgehog, the robin was taken on board by the Acclimatisation Societies in the mid-nineteenth century, and distributed around Australia, New Zealand, Canada and the USA. But to no avail: the robins failed to thrive and were never seen again. This was a good thing. In pretty much every case, introductions prove disastrous for the indigenous ecosystems. But it is an indication of the powerful affinity we have for the bird that, even if the species itself could not establish itself, the *idea* of the robin, and its name, did.

'Robin' is actually a friendly nickname. 'In Anglo Saxon,' Andrew explained, 'the bird was a "ruddock", and the early English called it "redbreast". It's interesting that the two birds with the best-known nicknames, Robin Redbreast and Jenny Wren, are also linked in folklore both as mates and – somewhat confusingly – as enemies.'

What is a robin, anyway? Well, it's not a close relative of the dunnock, despite the similarity of its Anglo-Saxon name. In fact the species behave in quite different ways. The dunnock often seems as much mouse as bird, scuttling under the bramble hedge. And, as my sparrow-man, Denis Summers-Smith, had told me, the dunnock is from the accentor family, while the robin is from the family of thrushes and chats – a family that also contains the nightingale.

Now, the nightingale might be a candidate for most-lovable beast. They've certainly received a fair amount of praise from the more tubercular poets. But Andrew was having none of it.

'They're just one-trick wonders,' he argued. 'Yes, their song might move great minds to great works. But the writers only ever write about the song. Now, the robin, well that attracts attention for a whole variety of reasons, which is why the robin is superior as a way of encouraging a deeper love of the natural world.'

To some people, indeed, the nightingale is a pest. A friend of mine was brought up deep in the forests of central Poland, where the nocturnal warbling of the resident nightingales would drive the family to distraction. For such a small bird, it is very loud. They stir great passions of love in moderate doses, but you only need look at what happens to people when exposed to beauty for too long and with no hope of release. I am sure that I would find myself becoming averse to Bach cello suites or Beethoven symphonies if they would not let me sleep.

And how many nightingales will perch on your fork-handle while you work in the garden?

This classic image, of the robin standing nearby to watch a gardener as he or she works, is not a fantasy but a regular occurrence. What have we done to warrant such attention? I was concerned that the story I had long peddled, about the robins looking at us as if we were pigs, was (sub)urban-myth. 'There is certainly some truth in that,' Andrew said. 'Robins will show a great deal of interest in freshly disturbed soil as it presents easy access to soil invertebrates. They're not as big and robust as, say, a blackbird. They're less able to excavate the ground themselves. So they have learned to rely on others. One of the best natural excavators is the pig. Robins will hop around in the branches as wild boar or domestic pigs bulldoze the ground, and pounce on the grubs. And, to them, we are just the same. As for the image of a robin on a fork handle, that's more than just the whimsy of the greetings-card industry. It perfectly illustrates their preferred approach to feeding. In the wild they'll find a low perch and use it to appraise the scene, before darting down to pick up prey. So in a garden, with the soil being turned and grubs and worms being exposed, well, it's no wonder the robin appears so bold.'

This got me thinking about the similarities between robins and hedgehogs: their shared boldness and the chance they offer us to get really close to a wild animal in our own gardens. Because they also share a preference for that particular habitat. Both are woodland-edge specialists, and, as such, benefited enormously from the arrival of farmers and, later, gardeners. The former cleared the forests, creating more of that woodland-edge habitat, and the latter then created further areas of suitable habitat in the form of a mosaic of hedges and flowerbeds, fruit trees and rockeries.

But our relationship with robins is more than just one based on our woodland-edge-dwelling porcine qualities. It turns out that the robins of Britain are a unique bunch. Nowhere else in the world does this relationship between people and robins exist. And this is not just some sentimental affectation; there is a tangible difference between our relationship and everywhere else in their range, from Western Siberia to North Africa. The robins of Britain are the tamest robins in the world. What accounts for this?

'Well, first there is the fact that we are a nation of gardeners like no other,' Andrew said. 'Ours is a crowded land and most of us need a little bit of nature close at hand, and because our gardens are perfect for the robin there is a lot of contact. But there is something more fundamental still. We *like* them.'

That took me a little by surprise. Somehow, 'like' wasn't a word I'd expected to hear in this context.

'I'm being serious. As a nation we've developed a fondness for robins that no other nation has, and it's a fondness that we have for no other bird,' he insisted. 'We like them because they are common and easy to see. We like them because they're tame. And we like them because they sing beautifully for pretty much the entire year. When things are dark, cold and hopeless, the liquid warble of a robin can bring a little brightness to our hearts. And also consider which garden bird it is that children first identify. The red breast makes them stand out, and they have a cuteness that children find endearing.'

I remained unconvinced. Surely we in Britain aren't alone in appreciating these qualities.

'There is, and always has been, a great difference between the way we consider our robins and the way it is seen by most other people,' Andrew explained. 'In Britain there's never been the sort of systematic persecution of songbirds, and robins in particular, that exists elsewhere. We may destroy them indirectly as we fragment the land and saturate it with agro-toxins, but that's not the same as the carnage that is culturally acceptable in continental Europe. And our love affair is far from a new thing. One of the earliest references I've found is from Chaucer, in "The Parliament of Fowls", from 1381. In that poem he presents the characters of the birds. So you get the false lapwing, the thieving chough, the vigilant goose and the wise raven, as well as the tame ruddock. Though the earliest-known mention is from around 800 years earlier, when St Kentigern revivified a robin killed by his fellow pupils.'

But what of the continent? It transpired that Andrew's use of the word 'carnage' was not hyperbole. Songbirds have long been on the menu in southern Europe. One estimate suggests that of the five billion or so birds that migrate across the continent each year, around one billion are deliberately killed. And that is in spite of the European Birds Directive that bans many of the more obscene practices. In times when food was short and starvation a real threat, liming branches to catch passing birds might have been defensible. But now, when starvation is a memory in Europe, the smearing of branches and posts with glue to catch resting migrants should be a cause of shame.

It feels so very strange to consider eating songbirds in general and robins in particular. Having held a robin I can warn potential gourmands that it won't make a filling meal. There is barely anything there beneath the fluff of feathers, other than a great big heart and a mellifluous syrinx. This fragility, Andrew thinks, makes it all the more wonderful when they do land on your hand.

We tried, briefly, to take our conversation out into the garden, to see if there were robins to be heard, but the intermittent rain combined with the noise of a neighbour's leaf-blower and Andrew's playful dog, Samson, forced our retreat back to the warmth of the kitchen. Where, I wondered, had Andrew's love affair with the robin begun?

'Both my parents have had a major role in this,' he began. 'My mother always had a tame robin who would feed from her hand. You know, I think there are far fewer people who do this now. Our love of the robin has become diluted a bit. With the possible exception of the magpie, people have become excited by *all* the birds of the garden. There was a time when we cared considerably more for the robin than any other garden bird.'

So how do I get a robin to perch on *my* finger?

'My mother had a very particular method,' Andrew explained. 'To start with she would put some food down on the ground – mealworms are great, but she used a bit of cheese. She would ignore it, leave it for the robin. Then, over the next few days she would get gradually closer until she would crouch down and put her hand on the ground. When the robin was taking the cheese with her there, she would begin to place the food closer to her hand. A few days of this and she would then put the cheese on the tips of her fingers. Always the robin would perch on the ground and peck the cheese from her fingers to begin with, then, as she moved it further back, well, she had to gauge it so that the robin did not just fly up and snatch it away. So, slowly, she moved it further up her fingers until the robin would walk onto her hand. She would then start to raise her hand off the ground. And when the robin found that nothing untoward happened, it would sit there. I've seen them singing from her hand. I've not tamed a robin for a few years now. One of the amazing things when a robin is standing on your hand is the knowledge that you would not even notice the effort required to extinguish its little flame of life. They are so small and so vulnerable and we are so big and so powerful. Yet who would do that? Who would hurt, or even capture, a robin? As Blake famously stated, "A Robin Redbreast in a cage puts all heaven in a rage". That apparent trust is very attractive, and common to both male and female, though it does seem it is the males that are bolder.'

I thought that males and females looked the same, so how can we tell the difference?

'You can't,' Andrew said.

Then how did your mother know it was the males that came to her hand?

'When they pair up and have mated, the females stop singing.'

Once you've done the training, I wondered, will the same robin come back, year on year? Do they remember?

'I'm sure they would if they could,' Andrew said. 'But the life expectancy of an adult robin is one of the shortest of all birds, only fifteen to sixteen months. They are beset by so many of the predators that find easy pickings in our gardens. Cats are only the most obvious; there are also sparrowhawks, magpies, crows, squirrels, nuthatches, weasels, rats . . . I could go on. Life for a robin is fraught. So it's a good job that a pair can have up to three broods of up to five young in a year. And as long as two of those fifteen chicks can survive, the population will be stable.'

And despite usually being fairly residential, they can move considerable distances. It is not clear why some robins choose to settle while others migrate, but it may simply be down to the need to acquire a winter territory. So each year we lose some robins to France, while gaining others from Scandinavia.

So why is Dr Andrew Lack, a specialist in pollination, such an advocate for robins? The reason is his father. David

Lack, the second director of the world-famous Edward Grey Institute at Oxford University, who so inspired my sparrow-man, is credited with developing the science of ornithology. He wrote *The Life of the Robin* during the Second World War and then *Robin Redbreast*, published in 1950. While the first of those books was purely natural history, the second was an anthology of poetry.

'My father loved *Robin Redbreast*, but it never quite sold as well as it should have,' Andrew began. 'And then, as my interest in poetry grew, so did my interest in this work. I found a great number of poems that had been written since then, as well as others he hadn't known about. So I suggested updating the book, and spent a wonderful time wallowing in the world of the robin.'

The new edition is a delight. It is also a fairly comprehensive revision of his father's book. Andrew shortened many of the longer prose selections and then added a large number of new and not-so-new poems. 'I've added about half as many poems again. The most obvious single example being John Clare. It's strange to think that he wasn't published in 1950 and that most of his very large body of poetry has only become available since 1980. Amazing that one of the most popular poets of our time was ignored for so long.'

There is a slightly childish part of me that fights against the popular, but I concede that Clare is one of my favourite poets, along with Ted Hughes. I have a collection of John Clare's poetry by my son's bed and when he needs some help to nod off, and I've lost interest in picture books, I reach for it and find that we both get something we need. I love reading poetry that 'wants' to be read, and he relaxes into the rhythm of my voice. But I wonder what has prompted the recent surge of interest in John Clare?

'I think it's to do with the times we live in,' said Andrew. 'We're seeing an awful lot of nature disappearing, and Clare was living through something similar. He was witnessing the enclosure and destruction of the countryside, and the end of a way of life that had endured for centuries, and it broke his heart. Similarly, we are seeing the fragmentation of what is left into unsustainable units that will eventually wither and die. And when that happens, something very important will also die in all of us.'

But the robin is safe, is it not?

'Oh, yes, it's rather nice to be so closely affiliated with an animal that isn't on its uppers,' Andrew agreed. 'It is a common bird. There are probably around five million in Britain and that number is currently fairly stable. They have not been affected by the great swathes of destruction meted out by industrial agriculture that has decimated populations of farmland birds like the grey partridge, the corn bunting, the skylark and the yellowhammer. And that's because the robin is not really a farmland bird. There will be a few bobbing around on farms, but their stronghold is with us, in suburbia.'

Is this habitual proximity alone enough to explain the great wealth of robin-inspired literature, folklore, art and poetry? Because, as Andrew has revealed in *Redbreast*, there is a vast range of material. What he has done is to take some of the key repeating themes and investigate them more deeply.

Perhaps the most widespread of the stories is the Saga of Cock Robin. Andrew recites the verse:

> '*Who killed Cock Robin?*
> *I, said the Sparrow,*
> *With my bow and arrow,*
> *I killed Cock Robin.*

'Despite being so well-known,' he says, 'we simply have no idea where it came from. The meanings have been lost in time. The oldest printed version dates from the early 1700s, so perhaps it might be tied in with the English Revolution. But I'm sure it's older than that. There is a fifteenth-century stained-glass window in Buckland Rectory that depicts a robin pierced with an arrow. But it could go way back into pre-literate pagan ceremonies.'

I was proud to be able to tell Andrew that I had an unusual connection to perhaps the most remarkable telling of the story. In 1861, the famous taxidermist Walter Potter set out a tableau of the Cock Robin story with no fewer than ninety-eight stuffed specimens. It became an object to which coach tours were taken, but eventually found its way into the hands of my friend and mentor, Pat Morris. Pat has written more scientific papers than anyone else about hedgehog ecology, biology, physiology and behaviour. But he is equally passion-ate about historical taxidermy, and when I visit him in Surrey, I have the pleasure of sharing the spare room with Walter Potter's tableau. I once asked Pat how much it was worth, given that the likes of artist Damien Hirst were known to be fans of the art of taxidermy. Pat mulled it over and concluded that he did not know how much it was worth, but if Hirst offered him a million pounds, he would say no.

The source of the rhyme is not the only subject up for

debate. Some have suggested that 'Who will toll the bell? / I, says the Bull' refers to a bull*finch*, which would, of course, fit into the bird theme. But perhaps the biggest area of controversy surrounds the view of ornithologists who have pointed out that the most likely culprit for the murder of Cock Robin is not a sparrow, but another Cock Robin.

'It's one of the points that threatens to tarnish the robin's glamour,' admits Andrew. 'I've seen people become quite upset when they discover robins behaving like, well, animals. The trust they seem to show towards us is a very attractive thing; it makes them seem like children. I'm sure this is one of the reasons why there are so many children's stories that feature the robin. But it's when they start behaving like birds rather than children that outrage can set in. Eating worms, for example, is frowned upon; but their biggest "crime" is murder. The males are very territorial and they will defend their territory with lethal force.'

And not all of the stories that follow the robin are cute. There's a regular association with death, though usually the robin's involvement is benign. Most persistent of these is the *Babes in the Wood* story, first recorded in 1595, where the two 'poor innocents' die in each other's arms:

> No burial this pretty pair
> From any man receives,
> Till Robin Redbreast piously
> Did cover them with leaves.

The robin's supposed habit of covering dead bodies with leaves or moss influenced other stories, and in due course the bodies in question were said frequently to be those of murdered men. Is there any explanation for this association?

'To begin with, I think this story actually emerged from an anti-Catholic rant. But as to where the detail came from, the only time you will see a robin with a beak full of leaves or moss is while it is making a nest. And as to why there is an association with death? Well, robins are very frequently associated with graveyards, and with good reason,' Andrew said. 'As we discussed earlier, robins are attracted by digging, and obviously there is a lot of that in a graveyard.'

Their association with the darker side of life extends to prisons and other places of incarceration, where robins have been seen, in Britain, as a source of comfort to the imprisoned. By contrast, in Germany Andrew found a story where they have been said to cause a prisoner to die of grief 'from hearing the thrilling sensations of loneliness conveyed in the song of the redbreast'.

Executed prisoners have also had well-reported interactions with robins. One individual made a nest in the skull of a hanging highwayman; another refused to leave the feet of a wrongfully hanged man.

Closer to home, the robin continues to have a strange and often unexplained relationship with death.

'When *Redbreast* came out I was lucky enough to be invited onto *Midweek* on BBC Radio 4, with Libby Purves,' said Andrew. (I had undergone the same nerve-wracking experience.) 'And before we went into the studio, while we were chatting, she told me a story that completely wrong-footed me. She explained that the night before her son died, a robin crashed into the window and broke its wing. None of them knew that a robin in the house was supposed to presage death. They tended it, but it died in the night.'

I wanted to check that Libby Purves was okay with my reporting the story in this book. In her reply to me, she added, 'On the day that my son Nicholas's book, *The Silence at the Song's End*, was posthumously published, another robin flew in, perched on the pile of books and, fearless, would not be shifted.'

Goosebumps crawled up my arms and down the back of my neck. But that was nothing compared to the reaction of a friend of mine, the wonderful artist Deborah Lomas. Knowing that she was interested in the apparently magical connections that exist in the world, I sent an early draft of this chapter to her to read.

'I thought I was alone in having a special connection with robins,' she told me. 'I read about Libby and burst into tears. My dear dad was a happy man, always whistling and singing as he worked in the garden. When he died, I tried to do the same. Working the soil comforted me, and every day a robin would visit me. I just knew that my father was somehow watching me through the little bird. One day I felt I was better, and told him so, and that was the last I saw of him.'

Where has this perceived link between the arrival of robins and the departure of souls come from?

'I think that part of this relationship stems from the great visibility of the robin in winter,' Andrew said. 'Like holly, they add a dash of colour to the bleakest months, and they also sing. In among the harshness of winter, until fairly recently, death would have been a very common visitor. Add to this the tameness of robins and the many stories of them being invited in to share a crumb when all is frozen, and it's not difficult to see how they would come to be associated with death.'

It's not just houses in which robins have been known to take refuge. There are a number of stories in which churches are used, and one in particular that neatly marries both strands of folklore.

'It is one of the most extraordinary stories,' Andrew said. 'When Queen Mary died in 1695 her body lay in state in Westminster Abbey for several days. During that time a robin took up residence. It was reported as the "Westminster Wonder" and taken as a great omen. Of course, the significance of the omen varied markedly according to the views of the writer.'

It is intriguing that a bird so associated with death should become the de-facto winner of the Birds-on-Christmas-Cards contest. And even before Christmas cards became popular, the bird featured on Valentine's cards. It seems that, while the winter presence and the lovely song partly account for this association with greetings cards, there is another, more prosaic, reason. The colour most associated with the Royal Mail is red: the vans, letterboxes and, for a time in the nineteenth century, the livery of the postmen – so much so that the postmen were nicknamed Robins. Then there is the familiar 'and finally' story regularly told at the end of springtime news bulletins, in which a village post box is put out of action for a few weeks as a pair of newly resident robins raise their brood.

'My favourite story, though,' Andrew said, 'is of the gardener who took off his coat in the morning, hung it in the shed, and returned at lunchtime to find a robin's nest in the pocket. Shows how fast they can build the little cup of leaves and moss.'

Stories of other unusual nesting spots were collected by Andrew's father, David Lack, and include kettles, beds, church lecterns, a cannonball hole in the mast of the *Victory* (Nelson's ship, though the nest was made after the mast had been retired to dry land) and even the engine of a working Second World War plane.

This ability to tolerate nearby human activity makes the robin the perfect urban bird. And its song, described by Emily Brontë as 'wildly tender', is one of the remedies for the grim grey of city life. Amazingly, the song can be heard at the strangest of times, with some robins taking it upon themselves to address the dead of night. Which has led to some notable moments of confusion: when a nightingale sang in Berkeley Square it might well have been a robin, and likewise, the Beatles might have been mistaken when they thought it was a blackbird that was singing in the dead of night.

'This tendency does seem to be mostly the result of artificial lighting, in cities,' said Andrew, 'but this is not the sole cause. The earliest reference I have found is from R.D. Blackmore's *Lorna Doone*, published in 1869, which contains the line, "Everyone knows that robins sing all night". And there are occasions when a robin will be excited into action by some other nocturnal disturbance, or even the song of the nightingale.'

'You know, our attitudes to the robin and its song have changed over time.' Andrew has made this observation after rummaging in the depths of the birdy archive. 'And I think

I have an explanation. When my father published his book he put the poems in seasonal order, starting in the autumn. But I decided to simply put the poems in chronological order, for the most part. Interestingly, it emerged that this fitted into the order of the seasons remarkably well. Now, there are exceptions; but the spring poems are the earliest poems, mostly Elizabethan. They celebrate the spring through the song of the robin, the reawakening of life as it moves towards Easter, and stories of resurrection. Interestingly, there have been attempts to squeeze the robin into the Christian story, with the suggestion that the red breast came about as a result of its attempts to pull the thorns from Christ's crown. As you move into the Romantic era, though, the most common season for robin poetry shifts to the autumn. This is at a time when people are being sucked out of the countryside and into emerging industrialisation, so there is a nostalgic feel: autumn is the time for nostalgia; the summer is gone. And, importantly, the winter is beginning to feel a little safer. Come the twentieth century, we can really celebrate winter. There is more food; we can store that food, and enjoy the season in a way most people would have found impossible just two hundred years ago.'

It is fascinating the way we can read shifting appreciations of the natural world in the rhythm of these poems. It shows how even this mundane, almost domestic, bird can link us directly to the wild and to our experiences of living as part of nature.

The Romantics were reacting against more than just the shift from a rural to an urban existence; they were looking at the whole shift in our perception of life represented by the rise of reductionist, scientific thought. There are no units for love and beauty. So where are we now? The war of attrition between the Romantics and the reductionists continues. Arts graduates make fools of themselves by talking nonsense about science in the media, and scientists likewise when dismissing the incalculable value of poetry. (And even in these scientific times, one myth continues to be perpetuated: that the robin has a red breast. If you look closely, it's clearly orange.)

Andrew is keenly aware of the importance of both views. 'Scientists do themselves a disservice if they deny the importance of the unquantifiable,' he says. 'Just look at love. People love birds; they love robins. Just look at the million-strong membership of the RSPB. Bird numbers may be falling, but if it was not for the love of a remarkably passionate constituency, they would have declined even further.'

I had looked forward to seeing my robin man because I enjoy his company. But at no point had I considered the robin to be a serious contender for victory in my fantastical little game. That was until we talked. We hadn't even seen a robin; there were no dramatic rides on motorboats; there was no crawling into tunnels; there were no soaking or freezing feet. There was no derring-do. Yet I was as excited by what Andrew had told me as I had been by the sight of leaping dolphins. Do robins have that special something? Are they the perfect foil for the hedgehog? One diurnal, one nocturnal? Both gatekeepers to the wild?

Who knows? There is still plenty more to see.

11. FOXES

'So familiar, yet so utterly wild'

When I was six, and first let loose on the world – allowed to play unsupervised in the fields behind my parents' house in Chester – my first encounter with a wild animal was with a fox.

I remember being sat in the ditch behind the broad hedge that once wandered as a border between fields, when I sensed someone, or something, looking at us. Perhaps it was girls – that was it, we were hiding from girls. And then I saw it, looking at us, probably wondering what on earth we were doing.

One year the farmer destroyed that amazing habitat. The oaks and ash that punctuated the route are still rooted to their spot, but each seems lonely: abandoned and slowly rotting. I still think of the flash of golden-red when I return to that place, and the feeling of those eyes focused on me. At prep school I produced an extra-curricular guide to the fox. And then Disney remade *Robin Hood* with foxes in all the most charismatic roles. (I think I fell a little in love with Maid Marian.) Foxes were important.

And as an ecologist interested in wildlife management you don't find many animals that rouse the same passions as the fox. Loved and loathed, persecuted and protected, the fox manages to get under our skin. This must be partly because it is so familiar in many ways, yet remains so utterly wild.

To find someone with strong feelings about foxes was not going to be difficult, but I was after more than that. I wanted to spend time with someone who knew the fox intimately. And I was quickly guided to Mike Towler. Some years back, Mike had found that a fox he had rescued was slightly brain damaged and therefore unfit for release. They became close, and the fox became tame, walking to heel like a domestic dog. But more than that, the fox also became a kind of passport. Mike found that walking the woodland with a fox at your side sends a signal that you are safe. And this enabled him to observe foxes and badgers unusually closely.

Unfortunately, when I got back in touch with him to book a date, his fox had died and another regular fox visitor had vanished. So I decided to call *the* fox expert, David Macdonald, Oxford University's first Professor of Wildlife Conservation and Founding Director of the world-famous WildCRU – the Wildlife and Conservation Research Unit. I had worked with him once, many years ago, studying the small-mammal populations around Wytham Woods near Oxford. This had introduced me to a work ethic like no other. David told me I would be doing three shifts each day, checking live traps and marking the mice, voles and shrews I caught, before resetting the traps. That was fine. What he had not told me was that the shifts were each eight hours long. Fortunately the daylight shift could be completed in around four hours and there was a sofa in the research station for me to collapse onto.

We had stayed in touch and I felt sure he would know someone perfect for the role of passionate fox advocate; someone who could explain to me why a fox was worthy of adorning my leg.

'It sounds like you're looking for me,' he said when I phoned.

David was tied up with international research groups and advising the government; I'd assumed he was well out of my league, and I didn't think he would be interested in something so trivial.

But he was, and after a little toing and froing to get our diaries to coincide, we agreed to give it a go. We arranged to meet at the appropriately named Fox Inn, right in the middle of David's old stomping ground of Boars Hill.

I knew it was his old stomping ground because David Macdonald had done something quite remarkable. He had turned his doctoral thesis, a document usually only read by three or four people, into a brilliantly accessible book, *Running with the Fox*. I had devoured it while doing my Masters and it had helped me to understand the importance of presenting complex ideas simply and entertainingly.

His book was full of irresistible anecdotes that leavened the hard science. For example, when he wanted to learn more about fox behaviour he hit upon the idea of taming a cub to the extent that it would allow him to follow it. What David had not anticipated was quite the level of devotion that this would require of him.

Warbling, as I now know, is what fox cubs do when they are lonely. For one so small and frail it is a powerfully expressive sound which, even after a short period of repetition, induces a great muscular tension in the chest of the listener. I lifted the waif into bed with me, achieving instant silence as she squirmed and nudged her way into the crevice between my collar bone and neck . . . Little did I know that this was to be where she would insist on sleeping for the next five months.

Punctual to the second, David pulled up in the car park, complete with battered Barbour jacket and boisterous Border Collie. I love dogs, a love that was reflected in those early fox encounters, I am sure, but when you're hoping to see wildlife, they can be a bit of a hindrance. But who was I to argue? I was just delighted to be out. The earliest hints of spring were appearing and a stiff breeze whipped the treetops into a dance.

Briskly stepping through a patch of what David described as rural-suburbia – big houses and gardens surrounded by extensive farmland, mainly horses and free-range chickens – we stopped by a gate and looked over the fields. 'This is where it all started, up here on Boars Hill,' David said. 'These fields here are called North and South Sands and were really the hub of my fox observations. What is special about this landscape is that it is at the interface between Oxford's urban foxes and the more rural foxes just a mile or so away.'

Fly, the dog, was leading the way, as I had feared, so we headed on into some woodland. As we did so I realised that I had not checked that my digital recorder was working, and in the brief moment that my head was down examining it, David hissed for me to look up – just in time to catch sight of the rump and brush of a beautiful fox, trotting across the path just twenty metres ahead of us. It was almost too good to be true. I lowered my gaze again – in time to completely miss the muntjac heading in the other direction. 'You would be surprised at the number of people who get confused by muntjac,' David said. 'The tendency is to think of deer as much larger animals, but the muntjac is not far off the size of a fox, and they also bark, which, to the untrained ear, can sound like a fox.'

This wasn't just a fox expert I was walking with, after all. David is also, among an intimidating number of other things, editor of the *Encyclopaedia of Mammals*.

'Hugh, we've been extraordinarily lucky,' he continued. 'This is one of the best places in the world to watch foxes, leaving aside some of the islands between Alaska and Russia, where they are so tame you can more or less stroke them.'

So are the arctic foxes generally tamer?

'I'm talking about our red fox. One of the reasons I started to work on foxes back in 1972 was because they are the most widely spread wild canid in the world, even more so than the wolf, which held that claim until we humans extirpated it quite so efficiently from much of its range. So there are red foxes, the species we just saw, all across Europe, Asia and America: from Alaska to the Gulf of Mexico; from the Arctic Circle to the deserts of the Middle East.'

This proves to be a fairly typical response from David, gently correcting me while at the same time feeding me answers to subsequent questions.

'Red foxes,' he went on, 'have made their home in every sort of habitat you can imagine, from jungle to concrete jungle. For a person interested in animal behaviour and ecology it is a source of fascination to study the ability of a single species to adapt to habitats so different that you might expect radically different specialisations.'

One such specialisation was rather close to my heart. Wildlife carers around the country had been telling me about an increased number of hedgehogs being brought to them having been attacked by foxes. There have been many stories about these two species, not the least well known of which is 'the fox knows many things, the hedgehog just one', and variants thereof. The essence of this ancient saying is that, while the fox is cunning, it cannot overcome the simple wisdom of the hedgehog. So, I wondered, did the increase in fox attacks on hedgehogs suggest that the fox was on the brink of overcoming a complication that had befuddled it since the time of the Ancient Greeks?

'I'm sure that many foxes eat dead hedgehogs scavenged from the road, but I'm not convinced that there's any upsurge in predation,' David said. 'I've seen many interactions and most of the time the spines of the hedgehog are sufficient to deter damage.'

I described the experience of a man I had met who had been asleep in a field when he was woken by a commotion. Bleary-eyed, he had watched as a fox sniffed around a curled-up hedgehog before cocking its leg, urinating on it and, as the hedgehog unrolled to escape the stench, grabbing and killing it.

'Did you ask him why he was asleep in a field?' David asked, with no little hint of disbelief. 'That's an old tale; both of these species are deeply folkloric. Many stories about foxes have been told and retold, and they're not always founded on truth.'

But I had a theory (I don't have many, and I was proud of this one). My theory was that the increase in the numbers of injured hedgehogs coming into wildlife hospitals stinking of fox was related to the shift from bin liners to wheelie bins. Urban foxes have been on the increase, in part, because of the easy food our wasteful society provides for them. A bin bag does not present much of an obstacle to a hungry fox, whereas the move to compulsory wheelie bins has created more of a challenge, and perhaps this has encouraged foxes to seek alternatives. Hedgehogs.

'I don't buy that at all,' David said, promptly bursting my bubble. 'Yes, there may be the odd encounter, but I would suggest that the smell of fox urine detected on injured hedgehogs has a more mundane explanation. Foxes are curious animals, and are likely to be fascinated by something as intriguing as a hedgehog. They will – I've seen them do it – mark things they find fascinating, including hedgehogs.'

I'm not so sure. While not as scientifically rigorous as David, many hedgehog carers I know have a good intuitive understanding of what is going on out there. I was reminded of an encounter I had witnessed, this time onscreen. While I was signing books at the Hay literary festival, a person in the queue presented me with a DVD that simply read 'Hedgehog and Fox' on the cover. I had seen a number of instances online in which people had set up cameras at hedgehog feeding stations and a fox had been seen snatching a careless leg and dragging away an unfortunate hog. So I was not expecting pleasant things as I slipped the DVD into my laptop back in my bed & breakfast that night.

The scene was familiar: fuzzy night-time CCTV footage of the goings-on around a feeding station. In the foreground was a bowl of food at which a fox was busily feeding. In the background a hedgehog appeared, meandering, as they will, towards the meaty dish. The suspense mounted as the hedgehog stopped to sniff. 'Run away,' I found myself saying to the screen. The hedgehog came closer, stopped again and I was by now really quite tense. Any minute, something horrible was going to happen, yet I was glued to the unfolding story. Again, onwards came the oblivious hedgehog. Any moment, the fox was going to lunge. The hedgehog paused again, milking the tension now. And then it happened. With a fierce leap and a bite, the hedgehog nipped the fox's rear leg. The fox jumped, startled, turned and ran. The hedgehog then calmly took command of the feeding station, and started to eat.

And with that tale I achieved a first: I had told the professor something new. He loved it, and this is what is so wonderful about him – not only is David Macdonald full of information, ideas and enthusiasm for his subject, he has an energy that would put most people forty years his junior to shame.

'My interest was never just academic,' he says. 'Throughout their range, foxes are involved in conflicts with people, and my interest in studying the animal was not only to understand them, but also, hopefully, to contribute towards the management and conservation of foxes.'

This was getting to the heart of the fox problem. While I had been won over by their beauty and wildness as a child, many people grow up believing that the fox is an outrider of the Four Horsemen of the Apocalypse.

'Some of the problems associated with foxes are pretty parochial,' David said. And I agree. The odd chicken or suckling pig could be considered meagre compensation for the impact we have had on the fox over the centuries. 'But there is one particular issue that is serious at an international level, and that is rabies.'

Again David has pre-empted my question. In the late 1970s, just as my interest in ecology was blooming, there was widespread fear of rabies. And yet this fear seemed to have subsided. Why were we no longer reading news stories about this terrifying disease?

'Epidemics of rabies have come and gone in Europe throughout recorded history,' said David. 'The most recent one started around the time of the Second World War in Poland, and began to spread westwards at a pretty constant rate, largely transmitted by foxes.' I was beginning to wish I had been on the receiving end of his lectures as an undergraduate. 'In the 1970s it looked as if the disease front would soon be hopping across the channel. Control had always been based on reducing the numbers of foxes by killing them, hoping that this would reduce the spread of the disease within fox populations because there were fewer foxes. Millions of pounds were spent and millions of foxes were killed, but with no effect on the speed at which the disease was spreading westward across Europe.'

David's hope was that if he learned enough about fox behaviour, he would be able to show the authorities why their schemes for control were not working, and come up with a more effective alternative.

And to do this, he had to learn from the very best teachers:

the animals themselves.

'But the problem was that I would only ever get fleeting glimpses, like the one we just had,' David said. 'And that would never be enough to fully understand even an individual fox, let alone the workings of fox society. I wondered about trailing a particular fox until it became tolerant of my presence. But rural foxes have undergone so many generations of persecution that they were too nervous. And then I thought about recruiting a spy. I decided to hand-rear a fox cub, win its trust without killing its wildness, and then follow it into the wild-fox world, taking notes.'

And this is why David ended up sleeping with a fox cub for five months.

'I had to call her Niff; as much as I think they are the most glorious animal, and as accustomed as I am to the aroma of fox, her constant presence in my bed did lend a certain extra herbal smell to my pillow.'

But wasn't there a risk that he would just raise a tame fox who provided an insight into the behaviour of *tame* foxes, rather than wild ones?

'That was a worry, of course – that Niff would grow up "de-foxed",' he replied. 'However, it turned out that I had reared the best possible teacher. That is not to discount any sentimental involvement on my part, but I always saw this as a privilege granted on vulpine terms, not imposed by me. And right now, let me make it clear, I did not create a "pet" fox. And it was not easy. Every fox cub I've known – and in the end I hand-reared more than twenty – has had a passion for both leather and electric cables. I've always found the lingering smell of fox urine not unpleasant, but I should warn anyone contemplating such a life that one landlady was unable to find another tenant for several months after my fox and I vacated the property.'

I can see the appeal of having daily contact with such a beautiful animal, of being allowed into its world. While radio-tracking hedgehogs, I experienced this to some extent, and began to see the individuals behind the species. This taught me two things: firstly, the fallacy of the Cartesian viewing animals as merely automata, without individual personalities; secondly, the ease with which one can tend towards the opposite extreme of anthropomorphising animals, imposing human characteristics on them. What David achieved, and I hope I did as well, was a balance between these two attitudes.

With Niff, David was able to begin to really unpick the workings of fox society. And his most important thing he learned was the reason for the fox's geographical success. Their ability to inhabit such an amazing diversity of habitats is because *fox society* is so diverse. For example, the size of the foxes' territories vary dramatically.

'We are standing within what I calculated to be the smallest territory ever recorded for a fox, just 18.6 hectares. Yet foxes I studied in the north of England had territories of 2,000 hectares,' David said. 'Now, this is not just of academic interest; it's directly linked to the work I was doing, looking at rabies control. What was evident was that a one-size-fits-all approach was unlikely to work and that control had to become more sophisticated. The idea that was current at the time saw the fox as an antisocial creature – especially compared to that other widespread canid, the wolf. Superficially, you could see why. Wolves would cooperate in bringing down prey bigger than themselves, while foxes eating mice and voles have little need to share. But what I was learning was that fox societies might, for example, include extended families that helped rear the young of a dominant female.'

But how was all this tied into rabies control? This was when David made an observation of great importance.

'Previously, people had killed what they thought of as individual foxes as part of rabies control,' he said. 'But I was finding that none of these animals really are just individuals; they are part of a social group. And when the social group is disrupted, the local status quo is thrown into turmoil. I worried that attempts to control rabies by killing individual foxes at random might actually be accelerating the disease's spread, due to what is known as social "perturbation".' David would go on to explain this theory. 'It led to an experiment being undertaken by colleagues of mine at the World Health Organisation.'

This proved to be a most valuable test of his ideas. Traditionally, the killing of foxes for rabies control had been done using poisoned bait. The new idea represented a radical change. Instead of killing the foxes, the new plan was to vaccinate them, leaving individuals alive and their societies 'unperturbed'. And the perfect place to do this trial was in Switzerland, where fox movements tended to be channelled through the rich mountain valleys. Imagine rabies spreading up a valley towards a point where the valley divides into two. At the point where one of these valleys forked off, air-dropped baits were dosed with poison as normal. At the mouth of the other branch, similar baits were dropped, but this time dosed with a vaccine to protect the foxes against rabies.

The results? Remarkably elegant: more than 80 per cent of the foxes consumed the bait in each valley. In the valley where the poisoned baits were dropped, around 80 per cent of the foxes were killed, but the disease continued to progress. In the valley that was vaccinated, the foxes were not killed and yet the disease stopped. This new approach, with its roots in the social lives of Oxfordshire's foxes, was taken up across Europe and now rabies has been largely eliminated.

'I named this the "perturbation hypothesis",' said David. 'The idea was that social turmoil among the survivors of killing campaigns could be counter-productive. In general terms the hypothesis says that if, for one reason or another, you kill an individual that is part of an intricate social structure, then you may be doing more than reducing the population by one; you may be changing the behaviour, and indeed the stress levels, of the survivors in that population.'

When there is a catastrophic outbreak of death within a fox society, the survivors may be thrown into social turmoil, travelling more widely and getting into territorial strife. And if the survivors are carrying rabies, this may actually accelerate the spread of the disease.

It's a theory that has a very contemporary relevance to the plans by various governments to try and control bovine tuberculosis. My badger man, Gareth Morgan, had made his concerns about this known to me most forcefully. David Macdonald has also been involved with the study of badger society, and identified exactly the same problems with perturbation. When badgers are culled, in other words, survivors are more likely to move beyond their stable territorial boundaries, and with that, more likely to spread the disease.

'One reason why attempts to limit bovine TB in herds of dairy cattle by killing badgers have not worked is because the rate of infection in cattle adjoining the culling area actually gets worse rather than better as a result – exactly as predicted by the perturbation hypothesis.' David spoke far more calmly than Gareth, but with deep seriousness.

'Of course, another reason killing badgers hasn't worked is that half or more of the herd infections are passed between cows, so tightening bio-security on farms is a priority,' David said. 'Nobody should underestimate the agonising choice faced by politicians. Bovine TB is a terrible blight on the farmers it affects, but the practicalities of killing sufficient badgers over a sufficiently large area to make a difference are daunting. On balance, I think a cull on a scale that is both feasible and socially acceptable is unlikely to help the farmers, and would do a lot of harm to the badgers.'

Arguments about badgers are set to continue for some time. Arguments about foxes, however, have subsided somewhat. When I was falling in love with foxes as a child, so much of the conversation about them was to do with hunting. Back then I simply had an instinctive dislike for the idea of taking pleasure from killing, and I remember many heated arguments with those who saw such sport as their right. It is with some satisfaction, therefore, that I look back at those arguments now that this 'fun' has been outlawed, and see that many of the arguments presented by advocates of hunting were utter nonsense.

At school there was a debate in which a hunter, arguing that these animals were virtually the spawn of Satan, stated that he had seen a pack of foxes bringing down a calf. Others argued that hunting was the only way to control the ravaging hordes of foxes that were rampaging across the countryside, killing chickens for pleasure and mugging elderly grannies for their pension.

But a little bit of reality undermines the image of blood-thirsty foxes. Quite clearly, foxes do not and cannot bring down calves. David Macdonald's research found that they had little interest in mutton, either. And while foxes will kill more than they can eat if they get into a chicken run, this says more about the chicken farmer than the fox. In the wild, chickens (or their ancestors) would take to the trees to avoid predators. The instinct of the fox is to hunt for what it can reach, which – in the wild – will most likely be those weak birds unable to make it to the trees. But if the entire flock is reduced to a state of weakness, both physiologically and in terms of their habitat, then the foxes will keep trying to get them.

So foxes are not 'evil', as some would have us believe. And the hunts knew that. Many of the larger ones used to feed and provide artificial shelter for foxes to ensure a plentiful supply of the 'pest'. Now the argument has shifted to the idea that if hunts are banned the very fabric of rural society will collapse or even that those objecting to animals being killed for pleasure are just conducting covert class warfare. It is hard to work out what to believe.

So how did David keep his work purely academic? How did he manage to retain professional distance? Was he able to keep the science 'in a box'?

Astute as ever, David decided to help me by rephrasing the question. 'I think you're asking me whether understanding fox biology is merely a matter of quantities, of measuring things. Of course, it is important to quantify what you do and make sure that the numbers add up and can withstand testing. But at the same time there is a lot about animal behaviour that is very hard to quantify. It becomes a matter of trying to get under the animal's skin.' And again, as with so many of my animal advocates, at the heart of the matter lies the issue of perspective. 'One should never underestimate the revolutionary importance of seeing the world from a different perspective,' someone clever once said. I think it was Herbert Marcuse.

'I used to take tremendous pride from the fact that, even as a young boy, I had some aptitude for dealing with and understanding animals,' said David. 'I tended to be quite good at taming them; particularly, as it turned out, foxes. I have spent many years of my life working directly with foxes and with their kith and kin – jackals and wolves, for example – throughout the world. I've worked on the great majority of species of wild canid that exist now. And I got to a position where I took a lot of pride in being able to guess what they would do next. Generally I was right. And that is not a quantitative skill. It might be broken down into numbers in some subtle way, but I think you could probably argue it was intuition. Not in the disparaging sense of "intangible and waffly"; intuition is actually about having some understanding of the rules that govern an animal's behaviour. I think there is a great delight in that.'

And I would agree with him. The pleasure you can gain from being able to enter, even slightly, into the mind of the animal you are watching is very special indeed.

But David was not done. He had seen through me. 'You

might have been trying to nudge me towards drawing some sort of contrast between science as something rather dry, and art as richer and more colourful. I deny the contrast absolutely. I think that understanding and appreciation are all part of the same thing.'

This struck me as important: one does not deny the other. It is a mistake to think that things retain their magic better if they aren't understood. Segregating art and science diminishes both. I remember a friend asking why I wasn't content just to watch a bird fly past, without trying to identify it. I went on at length about how, if it was an Arctic tern, it had already undertaken a mammoth journey to the far side of the globe, while if a black tip on a long red bill identified it as a *common* tern, it would have undertaken an enormous, if slightly less impressive, journey from the west coast of Africa. What I should have said was, 'Understanding and appreciation are both part of the same thing'.

'A true understanding of ecology,' said David, 'lies not just in the application of this line of thinking to populations of wildlife, but a far broader kind of thinking. It may sound like I'm just trying to justify my job, but what is required is a kind of thinking that has at its centre a consideration of how each ecosystem works. This sort of thinking should be applied to much more than just wildlife.'

An example of the complexity of people's relationships with nature comes from work David did looking at the economic impact of the presence of foxes on a farmer's land. If these foxes are taking some livestock (pheasants reared for shooting, say) this would impact negatively. But if the foxes were controlling rabbits, which otherwise would have eaten crops, then there could be an economic advantage to their presence.

We had been standing on the top of Boars Hill while Fly chased both his own tail and the scent of the many foxes that frequented the hill. The brisk breeze was eating through our shabby wax jackets, so we started a looping walk that would take us back to the pub.

I realised that I had started our conversation at the wrong end. We had spent most of our time talking the tough stuff of science, and I had still not asked two of the most important questions: why and how?

'Well, they're linked,' he said. 'When I was ten years old I found that I preferred the company of the wolves from Jack London's *Call of the Wild* and *White Fang* to the company of my contemporaries. So deep did this go that I wanted to be like a wolf. But the Surrey countryside was rather bereft of wolves and I found myself becoming intrigued with the wild canids that were present, foxes. At the age of eleven I wrote a sixteen-page book and managed to persuade my mother to sit out with me at night in a makeshift branch-and-bracken wigwam trying to see foxes. We failed. But I remember the first encounter incredibly clearly. We were watching the garden of a friend where scraps had been left out, and for about ten seconds I saw my first fox. This was seared into my memory even further,' he added, 'because at that moment a box of matches that my father was carrying in his trouser pocket spontaneously combusted, requiring a rapid dowsing with water.

'The "why",' he went on, 'is fairly straightforward. Studying the fox allows you to enter the best of many worlds.

There is still so much to learn. Right from the start, with Niff, foxes have illuminated my understanding of evolution and ecology. I get to see things that have never been reported before. It is applied work. At WildCRU we mainly see our role as one where we intervene to integrate better the lives of animals and the people they live alongside. Our job is conflict resolution, trying to work out what is happening and how it can be ameliorated. But there is another answer to the question, one that I wrote about over twenty years ago and that is still true to this day. I study foxes because I'm still awed by their extraordinary beauty, because they outwit me, because they keep the wind and rain on my face, and because they lead me to the satisfying solitude of the countryside.'

We were nearly back at the pub when he turned and said, with a glint in his eye, 'Of course, the real reason is because it's fun.'

And with that he whistled up Fly and I began my journey home, feeling more excited about foxes than I had in many years.

12. TOADS

'Like earth made animate'

This was never going to be the most straightforward of meetings. And while I had some idea of what I was letting myself in for, I obviously had not had enough of an idea. For a start, I had come with completely inadequate clothing. Of course, I knew it was going to be cold. It was January; the drive up into the Peak District had taken me through thick low cloud that bled into snow-covered fields, erasing all horizons. But I'd assumed we would be staying inside.

My advocate, Gordon MacLellan, had already donned intricately designed robes, before braiding his hair and neatly sewing the ends with waxed thread. He removed some of his more dangly earrings and then spent a brief moment painting his face with black lines, waves and spirals, before readying himself with another toasted crumpet and a glass of red wine.

I was shocked to see his bare feet, and not just because of the amazing tattooed menagerie that extended down to them from his legs. I had just been outside and the grass was freshly frozen into sparkling needles. The sky was black, pitted with innumerable stars. It was January, we were in the High Peaks and he was about to go outside, barefoot, in search of a very special toad.

I first met Gordon about twenty years ago; he was sitting in front of a tepee, drumming. This was having the desired effect of attracting a crowd. As an educational warden with the Mersey Valley Parks team in south Manchester, Gordon's job was to excite people, to get them to fall in love with nature. He was, and still is, very good at his job. That meeting started our friendship. Our mutual fascination with the wild world around us was spiced with memories of the time we had each spent in Africa. I still have the skull of a blue monkey he gave me, a highly prized treasure, which I allow poorly behaved child visitors to believe is the skull of a previous miscreant. As a disciplinary tool it works wonderfully.

And it was Gordon who got me thinking about tattoos. I remember his first. He proudly rolled up his left sleeve to reveal the outline of a toad creeping up the hairy jungle of his arm. His environmental art/celebration/education company is called Creeping Toad, so it was an appropriate image. But I had thought it was going to be coloured in. He blanched at the idea. 'I am never having another tattoo,' he declared. 'That was just too painful.'

The next time I saw him; there was still a hint of endorphin high shining in his eyes as he proudly showed me the beginnings of a rainforest higher up his arm. The addiction was established, and now he was developing a series of ecosystems on his arms, legs, back and chest.

It is clear from this, I hope, that Gordon is not your usual toad-enthusiast. Over the years we have had many late nights together, assisted by fine (and sometimes not so fine) wine. We share a love of nature; we share a love of *Buffy the Vampire Slayer* and silly sci-fi. We both work to inculcate people with a profound desire and respect for the natural world. Yet we come from very different worlds.

No, Gordon is not an alien. He is an extraordinary human who embraces radically different interpretations of the world we all share. Gordon is a Shaman.

I have met plenty of people who exist on the edge of delusion. The outer reaches of the environmental movement, for example, can attract some beautiful strangeness. And the world would be a lesser place without that beautiful strangeness. But Gordon is highly contradictory. His appearance would suggest familiarity with this rim of the known world, but as soon as he starts talking it becomes clear why he is booked up months and years in advance. His shamanic-inspired form of education is a winner, and there are thousands of children who enjoy a far more rounded life thanks to having spent time with him.

'The modern, public, face of the shaman is of the healer. But that is only a very small part of life as a shaman,' explained Gordon, while finishing the evening's preparations. 'When you actually look at it, and get past what makes good TV, you see that the shaman's role is to chart the relationship between their community and the world around them. This will mean anything from negotiating with the prey you hunt and kill, to working out the most harmonious place to pitch your tent or build a house. Healing is just one bit of this work. Healing is about the relationship between an individual and themselves, and between that individual and the spirit world around them. Shamans have a much wider function than that. I would argue that almost everything I do, regardless of whether it is in a school or in the wider community, has a shamanic function, because it is about the relationship between the community and its environment. My job is to get people to look at this relationship and help them see why it is so exciting. I have a little working motto that says "We live in a world that is worthy of celebration". That is the essence of my work.'

I sense that Gordon is keen to get on with his journey, but also keen to teach. 'You asked me about my connection with toads; that connection goes beyond the physical one – going out and watching them, or helping them across the road. The connection also has a spiritual element: I want to open myself up to toad spirits; I want to explore the *idea* of toad that lies behind all the little toads as they go hopping across the path.'

It is strange that, as he speaks, I am reminded of the poet Ted Hughes. The voice is different, but the desire to "explore the idea of toad" might come from one of Hughes's poems. I mentioned this and Gordon reminded me that he used to work with Ted Hughes, through the Sacred Earth Drama Trust. The more I read of Hughes's poetry, the more I think that he was, perhaps, as much shaman as poet. Gordon does not use poetry to get to his toad; his is a different, and more vigorous, path.

'Tonight's ceremony will involve dance – at least it will *probably* involve dance; you can never be too sure what will happen. I work with possession, I work by opening a door to the spirits to allow them in; dance is a way of becoming available to the spirits.' Not for the first time, I find myself slipping away from understanding. 'I live in a world where everything is alive and everything has the quality that we would call "spirit". This means that all things have the potential to communicate. And this awareness provides me with my inspiration and motivation. It also teaches me to listen. There is no judgement, you just listen. There is no hierarchy that says eagles and wolves are wonderful, and you can forget about the mice. The spirit of the mouse is just as important and just as powerful and just as wise. And the same goes for toads.'

At last, back with the toads. But as I am about to launch into my next question, Gordon stands up and it's obviously time to begin our journey to the toad.

I was under no illusions. The chances of dancing myself into a trance and stumbling upon Gordon's spiritual toad were remote. But I needed to try, needed to see what happened. And I needed far more clothing.

Just over a year ago I would have never even countenanced the idea of joining in a dance like this. But as part of my mid-life crisis, in November 2009 I succumbed to the persuasion of my dear friend Chloe to attend the 5 Rhythms class she teaches. As I had already done my first, and last, stand-up comedy routine, and had my first, and last, tattoo, why not do my first, and last, ecstatic dance?

Clearly this attempt to constrain my midlife crisis was a complete failure as I have been dancing almost every week since then. So I was not as completely terrified as I once would have been at the prospect of dancing. But while Gordon and his two dancing friends, Susan and Bryan, were suitably attired (aside from Gordon's bare feet), I was not. I put on as many clothes as I could find in my car, plus a sheep-skin hat. The grass crunched like gravel as we stepped into the dark of the Derbyshire night. Gordon crouched down to his knees, and I looked closely to see what ritualistic activity was taking place, only to find him plugging his iPod into its speaker-dock. The music began and all was quite peaceful. This was not the thumping dance music of Chloe's class; this was clearly meant as an accompaniment to the evening, rather than its driving force.

Delicate music wafted into dark that was so complete that, by stepping just a few paces from the lanterns Gordon had placed around the lawn, I was rendered instantly anonymous. This made me feel happier. I had felt very exposed, even in the dimness of the lantern-light. I realised how important the energising force of the music was to me. I spent my time in the darkness relaxing, meditating and wondering if I had brought any more clothes with me. Gordon soon retreated for his shoes, and soon I remembered that I had my tweed jacket in the boot of my car. There was very little opportunity for ecstatic writhing with all this on, but I did manage to calm myself, enjoy the darkness, the stars and the silhouettes of the trees, and the new moon hanging lazily above the scene. The candle by the small bubbling spring revealed a stream of slow water, thickening towards ice.

The idea of experiencing a new state of mind through dance is not completely alien to me. The state I arrive in after a couple of hours of dancing in a 5 Rhythms class, for instance, seems similar to that I've heard described by friends who have experience with stimulants such as Ecstasy. Perhaps there is a little of the puritan inside me: if I am going to have pleasure, I feel I should work for it. There is something undeniably wonderful about the flood of endorphins that rips through your body and mind when you dance, leaving you in love with the world.

None of that was happening out under the stars, however; at least, not for me. Gordon soon disappeared into the neighbouring field, his presence only betrayed by the clicking of the beads on his beautiful clothes.

In preparation for tonight, I had been reading a fascinating book called *Dancing in the Streets: A History of Collective Joy*, by Barbara Ehrenreich. She describes how even the most staid of modern religions began life as ecstatic sects, offering people direct connection with their god through dance and merriment. But the religions that became dominant sought to eradicate this fun, ensuring that everyone believed that the only way to god was through a priest of some description. I think early Christianity might have been fun; Dionysus-inspired revelry leavening a pacifist/anarchist cult.

Gordon's communion with the spirits was something similar to those early religions. He was not bound by predatory priests. He had sought, and found, his own connection.

After the dance we reconvened around the wood-burning stove and drank wine and told silly jokes. My favourite, and much repeated, being: what do you call a macho hippie? An 'alfalfa male'. I was still giggling as I slumped, exhausted, into my bed.

The next day dawned as crisp and blue as the night had been black. After breakfast, Gordon and I headed out for a walk to Pilsbury Castle, or rather the mound that remains from the eleventh-century Norman fortification. And this gave us the perfect opportunity to talk toad.

My first question felt almost like prying, but I had to ask whether last night he had danced with a toad. 'For me, Toad is always there. Toad is there as my friend.' When, I asked, did it all begin? 'I found toads and frogs as soon as I could walk, but I met Toad, as a presence, when I was ten. I'm now something unspeakable over fifty. I was brought up in a very sensible Church of Scotland family, but by the age of ten I had begun to question the language being used to describe my spirituality. It did not make sense, so I found words that gave me a different way of connecting with the world.'

So had this interest in toads always been spiritual? 'Oh, no, after school, instead of coming home with friends I came home with tadpoles in plastic tubs. As long as I can remember, I've been fascinated by amphibians. They're ancient and amazing. They first crawled out of the primeval swamps and took to the land over 300 million years ago. They were the first land animals, and it is them we must thank – nothing spiritual here, just evolutionary. It's from those earliest proto-toads that we evolved.'

Most of the 4,000-or-so amphibian species alive today are either frogs or toads; there are also a few hundred newts, salamanders and caecilians. They all have completely naked skin and lay eggs that need to be kept wet. In evolutionary terms they sit between the completely aquatic fish and the more terrestrial reptiles.

We were now zigzagging down a steep path towards the River Dove, trying to skirt the slippery ice and squelching mud. 'One of the things I love about the toads, and their amphibian brethren, is this ancientness,' Gordon went on. 'They pulled themselves out of the gloop of the mid-Palaeozoic and decided to stay as they were. They really have not changed very much in all that time, for the simple reason that their basic body form is one that really works.'

As we came to the bridge I heard a kingfisher, and we both glimpsed the darting flash of electric blue. The trees along the River Dove, up here in the High Peak, were low and intimate with the water. The branches appeared weighed down with the mass of lichen and moss. This was a very damp place. Ideal for toads?

'As toads are amphibians, they need water, or at least moisture, for their eggs to survive, but they are not as water-dependent as frogs,' Gordon said, as we leaned on the rail of the narrow bridge, the fast-flowing river below us. 'And here would be impossible for them to breed. They need still water – ponds, not rivers. Now, that is not to say that there are not hundreds of toads secreted in this valley right now.'

And right now, in early January, what are they doing?

'Not a lot,' Gordon said. 'But they do not hibernate, at least not technically. When hedgehogs hibernate they undergo a major shift in the way their body works, whereas frogs and toads just become more torpid. The main thing they do is hide.'

Frogs will often hide underwater, assisted by their amazing skin, which allows them to breath. Oxygen is absorbed through it and carbon dioxide excreted. But they cannot tolerate being frozen in ice, and also are at risk from suffocation if the surface of the body of water they are in freezes over for an extended period. This is because the ice traps gases from the decomposition of plants and prevents fresh oxygen from being picked up from the air.

Toads have a much drier skin than frogs and are not so good at this type of breathing, so they tend to head for other forms of sanctuary. They will bury themselves, digging down quite some way, or use the crevices in dry-stone walls or piles of fallen wood. So the beautiful stone wall we had followed up from the river could be harbouring a host of toads. That was a rather lovely thought. The wall was fascinating in its own right. For around thirty metres it was cushioned in moss, rich emerald green against the white stone; then, in the space of a metre, the moss faded into lichen and then there was nothing but stone.

To me these were obvious, and pleasing, signs of the river's impact on the surrounding habitat – moss requiring the river's dampness to thrive. But then, another fifty metres uphill, we encountered another patch of moss-laden wall. My theory was shattered. We both looked around, trying to see what was providing the damp conditions the moss required. Gordon spotted it first. This patch was adjacent to an old oak that was creaking skywards, and providing just enough shelter from the drying winds.

On level ground now, we headed towards the castle. In the woods to our right, Gordon told me, lay an excellent pond for toads. 'Not only is there loads of great habitat, but the farm buildings also provide shelter. And, in the spring, around late March, the toads will start to emerge, and over the course of a week, they will all have made their way to this and neighbouring ponds to make babies. Usually they go to the pond whence they came.'

This tendency to make for traditional ponds carries with it a considerable risk for the toad in these motorised times. 'I always carry a bucket in my car, next to the jack and the spare tyre,' Gordon explained. 'It's the best way to collect a mass of excitable toads and help them across the road.

And I'm not alone. There's a campaign run by the charity Froglife, "Toads on Roads". So far they've got around 750 toad-migratory crossings mapped. All over the country there are people who make a special effort in the amorous nights of early spring. They'll turn out with luminous jackets and buckets. There will be road signs and watchers stationed at either end of the toad crossing. It's not that dissimilar to a group of school children being shepherded across a road – only with more hopping.' To reinforce this, Gordon's car was adorned with a bumper sticker bearing the warning, 'I brake for frogs'.

The toads, on the other hand, emerge over just a few days, so can congregate in considerable numbers. And if there is a road in their way? Toads might have survived for 300 million years, but they do not react quickly to a new threat. They plod on and, if they are lucky, make it across or get scooped up and taken to their pond where the passions mount, as do the males.

The males have a particular grip they use to hold on to their beloved, called amplexus. There used to be an ad for the breath-freshener, Amplex – 'don't get a complex, get Amplex' was the punch line – aimed at creating panic and low self-esteem at the same time as offering a solution. There must be a link between this and toad sex. By the time that these fresh-breathed toads make it to the pond of their ancestors (or if none of those are available, any suitable body of water) the real business can begin. And this is where the croak plays such an important role. The pitch of the croak is related to the size of the male (the females are mute). Deeper equals bigger.

So when a male latches on to a female and another male approaches, a croak-off ensues. If their frequencies are similar, a tussle may begin. The female may be mute, but she is not passive. It has been found that if she is gripped by an unsatisfactorily small male, the female will move into parts of the pond with a higher concentration of toads, stimulating a bit of a clamour. This is not without risk. Sometimes it creates an avalanche of attention and all the males in the area join in the struggle, creating a writhing ball of aroused toad, with the female at the centre largely forgotten. Her death, by drowning, is not uncommon.

It is worth remembering that these are amphibians, which means that fertilisation is a little fishy; it takes place outside the body. As the female sheds her string of eggs the males uses his feet to feel what is going on and ejects sperm to coat them. This can go on for several hours, with short rest breaks for the amphibian equivalent of a cigarette. Frogs are rather less tantric about the whole thing, with the eggs erupting in just a few seconds.

Gordon's exposition on the sex life of the toad had taken us past the 300-year-old Pilsbury Grange, and then up to the castle. There were only ditches and mounds left, but excavations have revealed the extent of the wooden building, which incorporated the limestone knoll on which we now perched. The fact that this had once been coral reef made even the toad seem like a young species.

That they had survived for so long made the amphibians seem invulnerable. But would they be able to ride out the storm of humanity?

'Common toads are perhaps safer now than common

frogs,' Gordon said, as we settled down with a thermos of tea, admiring the view along the Dove Valley. 'And that's despite them being rather more choosy about their ponds. Toads have a capacity to hang on in there, they get by. They have resilience, in that they are long-lived. They will average between six and eight years, and can live up to twenty in captivity. But we shouldn't be complacent.'

As with all animals that produce a lot of eggs, most of these eggs do not survive the journey to become reproductive adults. The attrition begins immediately, with only a maximum of 5 per cent of the eggs making it into toadlets. And in the two to three years it takes to reach sexual maturity, 95 per cent die or are eaten. Only around five of the original 2,000 eggs that are laid make it through to adulthood.

'Frogs have been having a really rough time in recent years,' Gordon continued. 'Globally, numbers have been crashing at an alarming rate. Around one third of all amphibian species are threatened with extinction.' As with most life on earth, habitat loss is the main cause of this decline, but there is another problem that is wreaking havoc. 'There are hideous viral, bacterial and fungal diseases sweeping the globe. Here in the UK there were a number of unusual frog deaths that led to the Frog Mortality Project being set up with the Institute of Zoology in the late 1980s. What they found was that the frogs had been hit by a disease called ranavirus.'

Originating in North America, the disease spread rapidly. One of the reasons for this was, ironically, efforts by the general public to help frogs. By distributing frog spawn that was infected with the virus, they unwittingly spread it from one pond to another.

'Because toads do not spend so much time in water, they are less vulnerable to diseases carried in it. And also, their skin is tougher. So they have not been seriously affected by the virus. But they do have this strong connection with particular ponds, and this makes them very vulnerable to changes in the lie of the land.'

This vulnerability makes amphibians a useful barometer for the state of the natural world. Tropical frogs in Central America are being wiped out by changes in the climate, either being forced to climb ever higher up mountains to keep within their temperature requirements, or else breeding when there is no water due to erratic weather. While talking about this, I begin to feel that Gordon is using toad and frog almost interchangeably, and it forces me to reveal my ignorance and ask what the difference is between them.

'There is very little real difference between them, globally at least. Perhaps it's easier to go one step back up the taxonomic tree, to the 'tailless amphibians', also known as anurans. Here, in Britain, most anurans come from three species: the common frog, the common toad, and the natterjack toad. I've been ignoring the natterjack toad, mainly because you could write a book just on that one animal; and the difference between the common frog and the common toad is obvious, once you know what to look for.'

Natterjacks; I'd forgotten about these rare, sand-loving toads. I recall scurrying around the sand dunes while at prep school on the Wirral, avoiding the attentions of over-zealous teachers, while the natterjacks avoided the attentions of over-zealous school boys.

'Toads have shorter legs, and are drier, wartier and

rounder than frogs. I like to think of toads as round pebbles and frogs as pointy stones. But the main difference is in the way they behave. Frogs are hysterical. They will leap away at the first hint of danger. Toads will walk. And then only after sizing you up, perhaps even inflating slightly in an aggressive manner to warn you that a truly toxic brew of chemicals lies beneath that beautiful skin.'

Toxic? Beautiful? I had to take these in turn. Starting with the poison. Surely our toads are not poisonous? I'd thought that was the preserve of the brilliantly coloured tropical species.

'What do you think Shakespeare was on about in *Macbeth*? Why do you think the toad has been associated with witches?' Gordon is not impressed by my lack of knowledge. 'Watch a dog and a toad meeting. If the dog ignores the warning that the toad gives by inflating itself, and takes hold of it in its mouth, well, sit back and watch the show. Very quickly the dog will drop the toad and then it will begin to foam at the mouth. Those warts on the back of the toad are glands, which unleash a serious pharmaceutical arsenal.'

The sticky white substance excreted from the toad's back has four main components: chemicals related to adrenaline that cause muscular spasms; special toad alkaloids including bufotenine which can raise blood pressure and is also hallucinogenic; steroids, including bufogenin, that act as a local anaesthetic; and then the steroid esters, the bufotoxins. All in all, this is a mouthful that teaches a powerful lesson.

So was it true that licking certain toads could produce hallucinogenic states?

'Well, I've never licked a toad in pursuit of hallucinogenic delight,' Gordon said. 'But I have kissed a few in search of my prince! Actually, I have been poisoned, quite a few times, by toads. In one of my large tanks at home I used to keep a small colony of oriental fire-bellied toads. Cleaning the tank would mean siphoning out the contents and this would often result in me getting a mouthful of toad water. Within an hour or so it would hit. The first time I thought I had flu; very rapidly I began to shiver, and had a sense of not being quite with it. These days I know it's just a dose of toad. A very dilute dose, as well.'

The amphibian pharmacy is impressive. Poison dart frogs are not so named on a whim. The most impressively poisonous of these, the golden poison frog, packs enough poison to kill up to twenty people. But the species that causes the most trouble is the cane toad. Not only is this little beauty toxic as an adult; its tadpoles are toxic, too. Originally from South America, it has been deliberately spread around the world in a disastrous attempt to control the cane beetle, a pest of sugarcane plantations. And, to be honest, it is neither little nor beautiful. I defy even Gordon to be smitten by these galumphing sacks of pestilence and destruction.

'Oh! But *what* beauty; giant toads that look cast in bronze. They're exquisite, and enormous. Not that I am impressed by size alone, but a toad big enough to eat mice definitely has enchanting qualities.'

He is smitten, completely and utterly smitten, by toads. Perhaps a hint of the beauty in this beast can be found in Shakespeare, again – this time, *As You Like It*: 'Sweet are the uses of adversity, / Which, like the toad, ugly and venomous, / Wears yet a precious jewel in his head'. The bard was right about their toxicity, but what about that 'precious jewel'?

I've landed slap bang in the middle of Gordon's comfort zone, and I can feel him preparing his reply. But first we refill our mugs of tea from the thermos and break open some sweet flapjack, the oats just being an excuse, holding sugar and butter together.

'Should you try to kiss a toad, as you pucker up, look into its eyes.' It's instantly obvious why this man is such a hit as a storyteller. 'You talk about being nose-to-nose with a hedge-hog; well, try being nose-to-nose with a toad. It's not going anywhere. In fact it will try to hypnotise you with its magical golden eyes. These eyes feed into a whole set of toad law. Inside the head of a toad, according to the myth, is a stone, a magical stone. And if you can find a toad stone it will protect you from poison. Drop it into a drink, and if the stone fizzes, the drink is dangerous.'

For Gordon, though, the real magic of the toad lies not in the mythical stone, but in the very essence of the animal. 'Transformations are key to a toad's life. There is the obvious transformation from egg to tadpole to toad. And frogs do that too. But an adult frog is always going to be an animal. Toads are more versatile. They are like earth made animate. You might think you are looking at a lump of clay, when suddenly golden eyes will appear. They could be stones or pieces of wood that have come to life.'

These are the ideas and the stories that transfix children. 'I think the magic of the toad lies in that sense of transformation, of uncertainty, and the fact that they have been other things – eggs, tadpoles. This means that when we start to build stories with children, they can instantly start to challenge preconcep-tions. They will turn the toad into the hero, which is great, as there is such a pressure in the media's presentation of wild-life to fete the cute and fluffy. The toad is usually considered the outcast; it's an animal that carries with it many stories of portent or doom; it is not immediately cuddly. Yet, given the chance to gaze into those golden eyes, people are swiftly won over. They see that there, in their hands, is something magical, mysterious and, above all, most definitely wild.'

With that, we both begin to notice how cold we have become, sitting up on the castle, and begin the walk back to the cottage. The fire has been lit; we can see smoke from the chimney across the valley. And we have to get back soon. There is another night of dancing to be had. Perhaps this time I will meet my toad.

13. OTTERS

'Otter and water merge to make one perfect thing'

Some of the animal ambassadors refused to enter into active competition, but others took it very seriously. Huma was impressive with her bats, and Andrew wove his robin story with definite intent. Miriam Darlington started well. Her email came from 'mimdarling' and when we talked she was excited about a potential book deal and juggling getting children to school. It was like I had found a Devon version of me.

It felt like a good omen. It also felt like it was going to be easy. Easy in the sense that here was a very passionate, lucid and articulate advocate for a species that is simply one of the most charismatic on the planet. I was already beginning to think about how an otter would look as a tattoo, and had an idea of it throwing one of its impossibly fluid shapes around my ankle bone.

The otter was one of the first animals to leap into my consciousness, following fairly swiftly on from the fox. The stories of Gavin Maxwell left me craving contact with an otter. But I found *Tarka the Otter* too much. While, in retrospect, I realise that Henry Williamson's writing about the north Devon countryside is incomparable, I could not, at that age, get beyond the killing. And while it's true that those who hunted otters for fun probably knew more about their behaviour than anyone else, that seems small recompense for perpetuating this cruel and archaic hobby.

A few days before I was due to meet Mim in Totnes a package arrived. She had put together some homework for me. Two DVDs: one a documentary about otters in Devon shot so beautifully it was clearly the work of someone in love; the other David Cobham's 1979 film of *Tarka the Otter*. I watched the first with my boy Pip, who was off from school with a fever. We snuggled up under a pile of blankets and were transported to the lazy streams around the River Torridge, where the summer seemed more like the summers of yore, golden sun illuminating innumerable flitting insects, kingfishers zipping up and downstream like hummingbirds on speed – and otters playing. Pip was getting better, and by the end of the film we were wrestling like otter cubs in their holt. And *Tarka* was a bit much to share; that had to wait until the evening.

Our rendezvous at the otter sanctuary in Buckfastleigh promised a meeting with a captive otter. Driving across Dartmoor I felt a familiar urge to move to Devon. The bleak moor was studded with grey boulders of granite, like slothful sheep growing moss. And the valleys, with their lichen-bearded branches, carried roads like rivers, guided by banks of hedge and stone.

Of all the animals for which I've quested, this was the one I was least likely to sight in the wild. The only time I've seen wild otters was up in the Outer Hebrides, where I was thrilled by one boldly crossing the road in front of me before disappearing into the sea. But among the twisting roots of riverside trees along the River Dart, well, luck, or many patient hours, would be required. I did not have a large stock of hours. And luck, if my experience with the moths was anything to go by, was also in short supply.

Miriam was dropped off by her husband and she suggested that, as the sanctuary would not be open for a while, we take a walk down her favourite river, the Dart. Sometimes I've felt that it's taken a while to get close to my advocates; there has often been a bit of reserve, a few barriers and a general wariness that I might be more Louis Theroux than David Attenborough. I've always liked to think of myself as an unlikely offspring of the two, my thirst for the quirky spiced with a passionate love of nature.

But there was no hesitation when it came to Mim. Our shared passions for writing and wildlife erupted into a conversation that lasted all day, only to be interrupted by the formal interview I had to record. To begin with, it was a scattered and unfocused discussion, a simple revelling in the excitement of mutual delight. But it quickly focused down to the key areas of otter life: otter love and otter joy.

Walking along the bank of the Dart, we stopped every few paces to peer through our binoculars at the mud on the other side of the river, scanning for signs of footprints. While Devon is blessed with some of the best and most beautiful countryside in the world, I was beginning to wonder whether Mim was quite the bush-tracker I had taken her for, as we approached a bridge carrying a very busy dual-carriageway across the river.

We stopped under it to look at another sandy beach. Raising my voice above the traffic roar and pointing to the graffiti, I asked, 'Is this really the sort of place a self-respecting otter would want to spend any time?'

'When otter populations collapsed to the point of extinction, people began to think of them as terribly sensitive animals,' Miriam said. 'But they are not; they are tough, adaptable and resilient. Just look at their range. There are otters – what we think of as "ours" – from the west coast of Ireland to the far east of Russia, from Norway to North Africa.'

There are, in fact, thirteen species of otter spanning most of the globe, though it is absent from the Antipodes and Antarctica. But the only one we will see in the wild is the common, or Eurasian otter – *Lutra lutra*.

'Otters can be found in the most urban of settings, and here they don't mind the roads or the graffiti. They don't mind that it is not picturesque. And while the setting might affect *your* sensibilities, as long as the water is clean and there are fish and ducklings aplenty, well, it is not far from otter heaven.'

But what were the chances of us seeing a wild otter today?

'Probably around 0.1 per cent.' Mim smiled. 'They are pretty nocturnal, and they'll know we are here if they are nearby. And they could be nearby; you see that patch of trees over there? That's actually a small island, and is perfect otter habitat. We couldn't get there without a boat, so any otter would feel secure enough to curl up into a ball and sleep there.'

Are there reliable field signs that indicate where otters make their nests? I wondered.

'To be honest, when it comes to sleeping, adult otters do it just about anywhere, with little concern for creature comforts. I've seen them sleeping on the concrete steps of a boathouse. They're so well insulated by their fur that it's like they're carrying a fifteen-tog duvet wherever they go. In fact, there could be one sleeping in that coppiced willow. You could walk past an otter asleep at head height.'

So cryptic is the object of our desire that 0.1 per cent is probably about right when it comes to our chances of glimpsing one. Which makes the process of studying them all the more challenging.

'All it takes is a shift in perspective,' Miriam said. 'You need to rethink what you are classifying as success. Don't get hooked on just trying to see an otter; start getting excited by footprints and spraint.'

While we might know the otter as a rare and precious beast, the fact that a fairly well-known word – spraint – exists for its poop suggests that we have had far more to do with the species in the past. And that has been a misfortune for them in many ways. They have suffered due to the fine quality of their fur and the imagined threat they pose to fish, as well as providing entertainment for those who get a kick out of hunting and killing.

Spraint is wonderful, I've heard. There are people, Miriam among them, who wax lyrical about the stuff. 'Fragrant, tea-like with a hint of jasmine and fish,' she explained. Which is surprising when you consider the heritage of this mammal; it is a mustelid, closely related to weasels, mink, polecats and badgers, and only slightly more distantly related to the most famously fragrant of all mammals, the skunk.

Beyond its smell there is another reason to be excited by spraint. In fact, as one otter expert, James Williams, writes in his book, *The Otter*, 'We study the natural history of spraint, not of otters.'

'The spraint can tell us so much about an otter. What it has been eating, whether it is a male or female,' Mim said. 'It can also help us to work out where the edges of an otter's territory lie.'

Spraint is more than just the waste products of the otter's digestive system. It is also a vital communication tool for otters. Spraint surveys have allowed ecologists to begin to piece together the nature of territories. Fortunately for those tasked with trudging up and down the banks of rivers, otters tend to leave their messages conspicuously – on prominent rocks, logs or patches of grass.

For us, there was no such luck – no otters, no spraint and no footprints – so we wandered back to the sanctuary, chatting about the oncoming spring. Catkins were beginning to pulse from the ends of twigs and the range of greens was beginning to increase as the earth stirred from its hibernation. Nice to feel that an end was in sight.

We were met by David Field, who set up the sanctuary in 1984. 'I've always loved wildlife,' he said. 'But this is the result of a hobby that got a little out of control.' It was a Sunday morning and he was trying to finish redecorating the visitor centre, so we were left pretty much to ourselves. The first thing that caught our attention was the almost hysterical squeaking coming from the first pen on the right. 'Don't mind them,' Dave called, 'they just want their breakfast.'

We peered over the wooden fence to find that the squeaking and chattering was coming from two Asian short-clawed otters. These small and playful otters are a source of great confusion. Being far more sociable and diurnal that our nocturnal species, it is better for displaying to the public, and is the otter of choice for zoos and other wildlife parks. But this gives the public a distorted view as to the size of a 'real' otter, and results in reports of 'massive' otters being found dead on roads when, in fact, the otter in question is quite average in size for a Eurasian otter.

I didn't wish to be dismissive of the very obvious charms of these exhibitionists, but I had my heart set on more exclusive fare. Nearby, an otter named Sammy was housed. Sammy was a real otter, a Eurasian otter. And he is, without doubt, one of the most gorgeous examples of animal life I have ever seen close to. Watching him move around the enclosure perfectly illustrated the ability of this animal to inhabit two elements: land and water. His paws were soundless, like a cat, as he ran around looking for breakfast; he was hungry, too.

Miriam led me down a slope to a tunnel which provided an underwater view of Sammy's world. And as he entered the water we were transfixed by his sinuous, impossibly bone-less movement. Mim's hand rose to the glass, as if trying to touch him as he swam by; and in an instant he had turned. She quickly learned the game, and began to trace patterns on the glass for Sammy to follow with his nose. As he swam, his coat shed mercurial bubbles of air that wobbled upwards.

Miriam's look of utter delight as she leaned towards the glass and Sammy came to investigate her face was infectious. And I understood; years of infatuation fail to blunt such glee.

Sammy disappeared and we wandered back up to the surface and found out why. David was back, this time with a bucket of fish. He started to talk more about the work he did, and about the vital importance of education. He also told us that wild otters would come and visit, lured by ottery noises and smells. All the time, Sammy was standing up on his hind legs, begging. At least that's what it looked like. David threw a trout that arced through the air before landing with a splash in the large pool Sammy had been frolicking in. Sammy chased it like a dog. And then was gone.

It took a little time to work out what had happened. While the feline delicacy of his movements on land was delightful, and the grace of his swimming transfixing, it was that moment of transition that impressed me most. So often, it is the bit in between things that proves the importance of the things themselves. It was like the momentary silence after an orchestra finishes; or the space between the words in a poem; or the moment when a swing stops at the apex of its journey. It was as if the otter was made of water, his entry happened without ripple, splash or sound. They merged, otter and water, to make one perfect thing.

I had slipped into a reverie while watching this amazing animal and when my attention returned to David it sounded like he was describing this paragon of faunal virtue as a failure.

Aged only six weeks, he explained, Sammy had been found in Ireland, after his mother was killed by a car. He'd been taken to a wildlife-rescue centre where they had lavished care and attention on him. Too much care and attention, though, it turned out, because when he was old enough to be released, he was so used to being surrounded by children, goats, cats and dogs that to let him into the wild would have been dangerous. Dangerous for him, and for others – an otter's mood can turn as rapidly as its body in the water.

Anyone with a serious interest in rehabilitating injured wildlife would consider this a failure, however adorable Sammy might be. So now he had a new life, here in Devon. An important life. He had become, albeit involuntarily, an ambassador for his species.

Sammy's life as an ambassador was not perfect, though. His behaviour was showing signs of being affected by captivity. Wild otters wouldn't usually draw attention to themselves by standing on their rear legs and begging, preferring to remain close to the ground when feeding and moving around. This was a result of his living with humans who presented him with food from above. And then there was his tendency to run and swim the same route repeatedly. Before food arrived he would come to the front where we were standing, beg, drop back down to all fours, and then run to the back of the pen before diving into the pool, emerging and running along the same concrete wall back to us. This was not as depressing as the behaviour seen in larger animals at some unreconstructed zoos, but it indicated that Sammy was missing something. As were the American otters nearby, which had been so well fed on steak mince that they had lost the knack of the silent dive.

What was missing, in both cases, was wildness. When free to explore the world, an otter can travel great distances. The largest recorded territory encompassed 80 km of stream. And while they tend to live in a fairly linear habitat (along the waterway), they can stray surprisingly far from water. This is evidenced by the discovery of spraints in unusual places, as well as – all too often – the discovery of road casualties.

'While these animals are essentially solitary, their life is complicated,' Miriam explained. 'Usually if you see more than one otter together it will be a mother and her cub or cubs. For the females, adult males are useful only for mating; then they have to be kept away from the young. Cubs have to stay with their mother for longer than you might think, right up until they are the same size as her. In fact, the male cubs can get larger than their mothers, and yet she will still be in control. There is so much for them to learn. Size and strength count for a lot, but the skills required to thrive in the water take time to develop.'

Before we headed back to Miriam's, she wanted to take me to a nearby pub. This sounded like a fine way to spend a Sunday lunchtime, but as we got out of the car Miriam strode on purposefully past the front door to the back entrance to the kitchens. I followed, and we made a steep descent to the River Dart.

'I haven't been here for six months,' she said. 'But this cave is an absolutely fantastic place for finding signs of otters. Now, be careful here, it's very slippery. It would be a shame if you fell in and scared off an otter.'

Down at water level, we walked across moss-slick rocks until we came to a sandy beach and the mouth of a small cave.

'Okay, let's just stop here and have a look,' said Miriam, as she crouched down to peer across the wet sand. She pulled out a small torch and began flicking the beam of light across the surface, trying to see if it threw up any shadows. It didn't take long.

'Look, footprints. Come on, let's have a closer look.' Gingerly stepping onto the sand, wondering whether it might suck me in, I crouched beside Miriam. 'These are so small I wonder whether they might actually be cat,' she said.

Looking around, I noticed a small bundle of snowdrops beginning to peep through the thick moss. Then, on the ridge above us, I saw a most peculiar construction. About the size and shape of a military helmet, but made entirely of moss and grass, it was obviously some sort of nest. There was even a doorway. If I had been here with my children I would confidently have assured them that it was the work of pixies.

'Oh wow, you've found a dipper's nest,' exclaimed Mim. 'Now, come and look at these.'

A little bit further into the cave, Miriam was on her knees pointing excitedly at more footprints. 'It *is* otter. You see the small ones? Well, it can't be a cat . . .'

'Could it be mink?' I interrupted.

'No, definitely not mink. The footprint of a mink is star-shaped, with thin toes. An otter has its five toes in an arc. Now, if it was a cat, we wouldn't expect to see the claws as clearly as we can here. The only other animals that leave footprints that we could possibly confuse with the otter are the badger, the dog and the fox. The badger's footprint is much squarer, with all five toes sitting around the front pad, while in the case of the otter the heel is much narrower than the splayed front. A dog only has four claws, and a fox's footprint is roughly diamond-shaped.'

Miriam had almost entered a trance as she crept around on hands and knees in the dark, damp cave. Every now and then she called me over to examine fresh footprints; and then, more excitedly, to see some scratch marks.

'Rub your fingers on this,' she said, as she scratched a patch of exposed stone and lifted her fingers to her nose. 'Give it a good scratch and then sniff your fingers,' she instructed.

I did as I was told. I knew what Miriam was hoping for: the tell-tale scent of otter spraint. But as hard as I rubbed my fingers over the rock, and as deeply as I sniffed, I was unable to pick out the delicate aroma of otter poop.

'I think it might just have been otter wee,' she said, looking a little disconsolate. It is rare to be disappointed to find only the *urine* of an animal on your fingers, not its faeces. Miriam disappeared further into the cave. She was crouched, in awe, over a puddle of water. 'This is quite simply the most perfect footprint I have ever seen,' she said, pointing into the water. And it was perfect: pristine, sharp-edged and completely clear. An otter's footprint. 'The water must have remained completely still since the print was left. But I'm puzzled at the lack of spraint. The only explanation I can think of is that this cave has been used by a mother otter and her young.'

For an animal that uses its faeces as a very deliberate method of communication, when the time comes to be 'silent' waste must be disposed of discreetly. When caring for vulnerable young, female otters are thought to defecate directly into running water.

Miriam was beginning to back out of the cave. She never went further than this. I share with her a very logical dislike of dark, enclosed spaces and can only marvel at what drives cavers to such extremes. However, I felt brave enough to try going a little further. Crouching right over, avoiding the jagged limestone ceiling, I shuffled further into the planet. After wading through water that threatened the integrity of my Wellington boots, I reached another beach of sand. By now, the only way to guarantee not smashing my skull on the ceiling was to spend as much time as possible lying on my front. The damp permeating my clothes was increased by the water dripping from the ceiling. I could make out a pile of something in front of me. I might have struck gold. But I realised I'd left the torch with Miriam.

Fortunately, modern mobile phones are a little like electronic Swiss-Army penknives. Flicking through the 'blades' on my phone I found the torch. Shielding it from the persistent drips, I shone the light on the pile ahead of me. Why anyone would want to build a small sandcastle in the depths of a cave was a mystery. But that is what I seemed to have found. A small and very deliberate pyramidal sandcastle. I shouted back to Miriam, who seemed remarkably unperturbed. 'Yet more signs of otters,' she said. 'They will often build small structures on which to spraint. Have a sniff, what's the smell like?'

I'm glad the entire exchange was recorded for posterity because later, when I thought back about my time in the cave, I did wonder whether I really was crawling around in my nice tweed jacket sniffing sandcastles in the hope of catching a whiff of otter. Again, nothing, not even the slightest hint of tea, jasmine or fish. But beyond the small sandcastle I caught sight of something brown and sticky – 'sticky' in that it looked a bit like a stick. I called back to Miriam, saying I had found otter poop. 'Is it crumbly and blackish?' she shouted. 'Bring it out with you; we might be able to see what it's been eating.'

I have two children; I have walked dogs through the park. I am well versed in dealing with the output of others. But there is usually some sort of protective barrier, a plastic bag or wipes. Thus I was rather hesitant when confronted by this exciting turd. Steeling myself, I reached forward and picked it up. 'What does it smell of?' came the call from the mouth of the cave. I brought my prize to my nose, and inhaled deeply.

There is no other way I can think of describing it; perhaps I lack the necessary olfactory vocabulary: it simply smelled of poo. I relayed my discovery to Miriam, who suggested that it might be human. I dropped it, only to hear her giggling. 'It's probably fox,' she said. There was only one course of action open to me, and I propelled the turd with some force towards my tormentor.

I could see further into the cave from this little beach, and I was very tempted to go exploring. But I had a few conflicts going on, not least of which was the saturated state of my clothing. So I turned around and began crawling back into the light. By the time the ceiling had lifted enough for us to squat together, I could see that Miriam was still enjoying my discomfort. I really needed to wash my hands, and luckily there was plenty of water available.

Finally able to stand upright I stretched and suggested that we make use of the pub for – at least – a cup of coffee. 'Do you really think they'll let you in looking like that?' Miriam said. I looked down at myself. While the tweed jacket had proved its worth, my trousers were in a state, wet and thickly coated in sandy mud. But I was able to scrub up enough to gain entrance to the pub, where I found a welcome cup of coffee and, best of all, soap.

Back at Miriam's home, on a hill overlooking the outskirts

of Totnes, we settled into her study, while her dog Barney rested his head in my lap.

Miriam began with her beginnings, and opened up a box-file filled with issues of a publication called *Otter News*. Issue One, Summer 1977, was handed carefully to me. As I started to turn the pages I was whisked back to being ten years old, when I collected every single piece of information I could about foxes, and wrote a newsletter for the benefit of no one but myself. However, Miriam had gone much further. Here were newspaper cuttings, photographs, even some editorial comment – all collected together in an otter scrapbook. And she didn't do it just the once. 'I kept it up for four or five years, until I discovered something else. Boys.

'I produced the first one when I was eleven,' she went on. 'I had just read *Tarka the Otter*, and I was basically living my life as an otter – without all the water and fish, of course. My heart was the heart of an otter and the more I learned about the animal, the more I felt like someone who had discovered the true meaning of love, only to discover that the object of that love had gone. I fell in love with otters when they were on the brink of extinction in this country.'

This was not hyperbole. In the late 1970s there were perhaps 200 otters left in England. For centuries, otters had managed to survive the depredations of huntsmen. But the enthusiasm with which agrochemicals were embraced in the 1960s, as an easy way of increasing farming yields, swiftly brought this sublime animal to the brink.

The sort of surveys which form such an important part of ecological research these days were not common in the middle of the twentieth century. In retrospect it's all too easy to see the missed opportunities. But in the early 1960s, when concern about the state of otter populations was becoming more widespread, many different factors were thought to be to blame. Was hunting responsible? Was the canalisation of waterways the problem? Escaped mink were also considered to be a disease threat.

Now we know the cause of the decline: the toxic brew of organochlorides used in industrial farming, such as dieldrin, aldrin and heptachlor. Brilliant at killing insect pests, these chemicals were widely used as a seed dressing or a sheep dip. What is perhaps surprising is that the link was not made between the decline in otter numbers and the earlier decline in bird populations that led American biologist Rachel Carson to publish her groundbreaking book, *Silent Spring*, in 1962. While her work focused on the impact of DDT, these newer chemicals were related.

By 1968 Britain had banned these organochlorides in dips, because their use had been shown to affect immunity and reproductive success in sheep. But the link was not made to otters, because the timing did not seem to fit: otter populations continued to decline even after these chemicals were prohibited. The assumption was that if chemicals were the cause of the declining otter population, banning them should result in a resurgence in numbers.

Again, with the benefit of hindsight we can understand what was going on. These chemicals are fat-soluble and remain stored in the bodies of animals exposed to them. Because they are not excreted, a concentrating effect occurs from the bottom of the food chain to the top, which means that the fatty fish most favoured by the otter, the eel, delivered a toxic punch. But there was a delay in this effect, which is what confused people. Organochlorides had been banned, but the chemical contaminants were still in the ecosystem, continuing to be concentrated up the food chain towards the eels.

I find it hard to understand how those who enjoyed hunting and killing these beautiful animals for fun could have continued until 1976, when their prey was on the brink of extinction in this country. Miriam was well aware of the hunters' world when she was writing the Editorial Comment in her delightful publications. I can just imagine the 11-year-old girl, unrestrained by adulthood, honest as only a child can be, as she wrote 'Otter hunting is a bloodthirsty cruel sport'.

Otter hunting may have been banned, but the depressing news is that we are killing as many otters each year on the roads as were killed by the hunters. The atmosphere in the room had changed; we had both been subdued by this catalogue of abuse. So it was a relief when Miriam opened up another edition of *Otter News* and took us back to her all-consuming passion.

'I wasn't distracted by boys for that long, and quickly came back to the otters; well, in fact, that passion never really went away. I stopped producing new issues of *Otter News*, but I did go back to the earlier ones and, with a friend, started to add illustrations.' Miriam handed me the scrapbook. One of the drawings looked like a hedgehog. 'It's meant to be an otter!' Miriam said with faux indignation. '. . . Okay, maybe it's a slightly fat otter.' But what about those spines? 'No no, that's wet *fur*.'

One of the things I loved about Miriam's passion was that,

at heart, she was a poet, not a scientist. Therefore our conversation was very rarely linear.

'The fur of an otter is a remarkable thing,' she said. 'The sea otter's coat has more hairs per square centimetre than humans have on their entire head. When people found how amazing this fur was, well, you can imagine what happened.'

Miriam did not need to say more. The history of humanity's attempts to manage wildlife through the marketplace is littered with abject failures. The sea otter was once common along the Pacific coast, Northeast Asia and North America. Within just a few decades of exploitation beginning, these amazing animals were brought to the brink of extinction. Legal protection was introduced as early as 1911, which allowed numbers to recover. But they still face many threats.

'The amazing complexity of the ecosystem in which the sea otter lives feels almost poetic,' Miriam explained. 'It really makes you understand how interlinked everything is. At the simple level, sea otters eat a lot of sea urchins, and by keeping the number of sea urchins low, they allow the great beds of kelp to flourish. These kelp beds play a vital role in ensuring a rich diversity of food for the otters. But there is another cycle, which we humans have influenced. It now seems that killer whales are hunting sea otters, to such an extent that the population is again under threat. You might think this is a natural battle, so surely it should balance out. But it is *not* natural. Intensive whaling removed a lot of the killer whale's natural food sources – small whales, or young whales. So they took to feeding on seals and sea lions; and when these too declined, the killer whales turned to the otter.'

I have to keep reminding myself that the PhD Miriam is

doing at Exeter University is not in its Zoology department but in its English department. Her passion has led her to read a great deal. And also to experience a great deal. 'Just after I finished my degree – I would have been twenty – I bought a campervan and headed off to Scotland. It was a mad and obsessive departure,' she said. 'Until then, I had lived chunks of my life as an otter, had read obsessively about otters and had seen captive otters; but I had never seen a wild otter. And I had to. I remember it so clearly. Having parked in the dark by the shore on the West Coast of Scotland, I was woken at four o'clock in the morning by an enormous crunching sound. It was light, midsummer. I quietly opened the door and looked down to see an otter eating a crab. You saw how they were eating at the sanctuary earlier – noisily, messily? Well, this otter was doing the same, holding a large crab between its dextrous paws and ripping it limb from limb.

'A friend suggested I might need treatment for my obsession.' Miriam chuckled. 'But I cannot think of a more perfect cure for malaise than time spent in the company of this remarkable animal.'

I'd met ambassadors who had talked about wanting to get under the skin of their chosen animal. But Miriam had taken that idea a step further than most. 'I'm incredibly curious,' she said. 'I tend to follow up every lead as it presents itself. So when I had an opportunity to learn how to track otters, I jumped at the chance, and now I know how to spot otter trails in the morning dew, for instance. When I got drawn into the ecology of the otter, I read everything I could. And when I found the Otter Project, based at the University in Cardiff, where dead otters from all over the country are sent to be analysed for evidence of pollution, well, I had to go and see.'

But this wasn't an easy visit. 'I was incredibly nervous; I simply did not know how I would react when I witnessed a post-mortem. I was given a job as an assistant, initially, taking notes and weighing organs. Having something specific to do helped; if I had just been watching I think I would have been sick, or burst into tears. And they needed help; they had a backlog of over a hundred otters that needed to be checked. I remember the first one most clearly. The scientist unzipped the otter using a scalpel, and we marvelled at the quality of its muscles. This had been a very fit animal, before it was hit by a car. I was handed a slice of liver, then a bit of muscle tissue. Then the gallbladder, followed by the adrenal glands. Each of these pieces was weighed. And then she handed me the heart of the otter. It was about the size of a satsuma; very red. It struck me that this was possibly the closest I would ever get to a wild otter, and here I was, holding its heart. I became absorbed, amazed at the plum-coloured beauty of the workings of this otter, glistening like the inside of a flower. And then it was over. At the time, I didn't realise quite what an effect it had all had on me.'

So greatly was Miriam affected by the experience that she got on the wrong train and found herself in Gloucester rather than Exeter, at which point she felt tendrils of panic creeping around her heart. 'I managed to hold myself together until I got home, three hours later than I expected, and then I collapsed into my husband's arms. I couldn't shake the smell of decomposing otter; it stayed with me for weeks.'

Miriam was an amazing mix of open-hearted romantic and

rigorous accumulator of scientific knowledge. And as such she was a powerful advocate for the otter. But how safe is the species now? Hunted for pleasure, poisoned by pesticides, the otter has had a pretty rough deal. But it felt like the worst was over.

'They live on a knife edge,' Miriam said, deeply serious. 'Even without human influence, they struggle to maintain their energy levels. They need to eat an enormous amount of food every day just to give them the energy to keep warm when swimming through cold water, in order to feed. And, in the past, we have made that impossible. Now they are hoiking themselves back into balance. I think the government and the Environment Agency would like us to believe that this is a massive success story. In some ways it is. The otter is not extinct. But it seems that lessons are never learned. After the organochloride poisons there came the organophosphate poisons, which have been followed up with synthetic pyrethroids. And these are just the chemicals that are intended to be toxic. Who knows what impact the gradual decay of our fireproof society will have. There is flame retardant in so much of everything we buy these days. And on top of all this is something that nobody seems to be checking at all; and that is the synergistic effects. The dose of flame retardants on its own might leave an otter unharmed, but what about when this happens to an otter already exposed to a dose of agro-toxins? We simply do not know enough about what we are pouring into our waterways. And the sad thing is that the first we notice will be when the otters start to vanish.'

Even if the otters do return to their former strength will

they be secure? New laws have been introduced, banning the hunting of mammals with dogs, but laws can change. And the hunting world clings bitterly to its anachronistic view of wildlife. Miriam's current project is intended, in part, to prevent a resumption of hunting. She is writing a book, *Otter Country*, which I am sure will seduce readers into otter-love. Her joy is infectious and her book will put her in the company of some of our finest nature writers, adding another beautiful contribution to the literary legacy of this poetic animal.

The Invention of Otter

No one can say how
it came out of the water
or how it plucked pebbles
from the river's pockets
and made thoughts

no one can say
how the night made nostrils
and whiskered its way
from the roots of an oak

no one can say
how its rudder thickened
with the wind, made fur
ripple into a stream
or how the storm muscled
a heart out of the moor

who can say when the eel
learnt fear, or the trout
first felt speed shiver
through its sides?

    Only, when it swims
all the water leans toward it
frays air into a swilling of purls
and streams love themselves
more deeply

    all I can say is
at that moment, poetry
nosed its way into the world
took its place among the four elements
made them five
    and now
when night silks the water
the weave of it says *hush*
keep it secret.

(Poem reproduced by kind permission of Miriam Darlington.)

14. BEAVERS

'Sitting upright on its hind legs, its flat, hairless tail acting as a support'

Beavers were a bit of a joke. I had included them in my original list of animals after a friend of mine had mentioned that her sister-in-law's brother-in-law, a man named Paul Ramsay, had released some into his estate in Scotland. I sent an exploratory email, heard nothing and returned to the task of tracking down advocates for British wildlife species. After all, beavers are extinct in Britain.

But then I was chatting to my next-door neighbour about this epic journey around the country and he said I really must talk to his friend's father, who was passionate about beavers. So I took down the friend's father's details and found out his name: none other than Paul Ramsay.

This time I got an enthusiastic response and an invitation to come and learn about his project to reinstate beavers as a healthy component of the national fauna of Scotland. Paul was keen to see how these free-spirited engineers might assist his quest to improve the biodiversity of his land.

All well and good. But I had no idea who or what I was letting myself in for. Revelation came from the *Daily Mail*. I am not in the habit of reading this particular paper, but I had come across it in a waiting room. Flicking through the hyperactive banality, I found a story about a protest my wife had been filming. One of the large anti-cuts demonstrations, in London, had resulted in a band of very peaceful campaigners temporarily occupying Fortnum & Mason as they raised awareness about the efficiency with which it manages to avoid paying too much tax. And one of the many was Adam Ramsay, son of my beaver man. The peculiar piece went on to suggest that the splendid store 'must have seemed like a home from home . . . Ramsay, 25, hails from a remarkably privileged background . . . The family seat is Bamff House' – described by the paper as a 'small castle'.

How exciting. A castle – though the picture the *Mail* published showed a rather dishevelled vision that might charitably have been described as displaying a rather faded splendour. All the same, a castle. I was wondering how Paul might be coping with his son's incarceration, albeit for just one night.

The journey up to the beavers of Bamff, north of Dundee, was a long one. But it was wonderful to see time run backwards, as the train ran forwards. It was mid-April, and we caught up with the blooming daffodils in Cumbria. My daffs in Oxford had long since browned, but these were in their prime. Apparently spring moves north at walking pace, and goes slower up the hills, of course.

The meditative qualities of a good train journey, once I had claimed a seat, were deeply therapeutic. I was soothed by the passing scenery: the splash of blue along a railway embankment as once-domestic forget-me-nots broke for freedom; the wave of yellow as buttercups formed a phalanx advancing into a meadow; the elbowed thorn trees walking on their own across a field – all this made my heart sing. I had been spending too much time in my shed, writing.

From Dundee I got the bus to Alyth, where I rang the number Paul had given me. Paul's daughter, Sophie, answered, and told me that her mother, Louise, was coming to find me. 'She's in a silver Prius and is very jolly; oh, and could you pick up some milk, a few tins of tomatoes and some onions?'

Sophie's instant ease and informality made me think I was going to settle in just fine. And sure enough, Louise proved to be as jolly as her daughter had promised. It turned out that they had been highly amused by the intrusion of the *Daily Mail* into their lives; in fact, she said, the piece was to be given place of honour in the toilet as soon as time allowed.

Turning off the road onto a rutted track, we bounced through dense woodland broken by the occasional large pond. Rounding the last corner I was rather surprised to see what a poor job the picture researchers at the *Mail* had done. They had chosen the one image of Bamff House that would elicit sympathy. The view before me caused me to gasp. The building was stunning. Yes, it needed some attention, but I have always been a sucker for the faded look. While there are bits that date back to the sixteenth century, the more magnificent additions came in the mid-nineteenth. There was even a tower.

Paul was in the kitchen and, after we had had a cup of tea, quickly got down to business. 'Look,' he said, 'we'll go for a walk. I think it's better to show you what I've been doing.' Equipped with binoculars, walking stick and sturdy boots, Paul led me into the 1,300 acres of land that he is in the process of transforming. Or rather, that his rodent friends are in the process of transforming. He strode purposefully past the topiary, telling me how the hedge revealed the story of the origins of the house of Bamff. Way back in the thirteenth century, the first recorded Ramsay of Bamff made a magical remedy from a stew of diced white snake to cure the King of Scotland of a large hair-ball, and in gratitude he was gifted this fine land in Perthshire.

As yet, no beavers had carved into the hedge. But I was sure it wouldn't be long before the topiary griffins were joined by a species a little more lively, and a little more local. But could beavers really be described as local? Reintroducing species to a locality where they have become extinct raises some interesting questions. Was it wise? How long until they become officially 'resident'? We headed out along a small stream that was little more than a drainage ditch, with Paul using his walking stick to help paint a picture of the history of both the land and himself.

Originally he had intended to manage stock, as many generations before had. Sheep and cattle, mainly. But he also had a deeper interest in the landscape, an interest that remained hidden to him until he attended a conference about Scottish environmental history, in 1997. 'They were talking about species that had become extinct. I'd never thought of the beaver as an indigenous species, and here they were describing how it could be a valuable tool in improving woodland biodiversity, as well as river and wetland restoration. At the end of the discussion I got up and said, "Look, what are we waiting for?"'

What Paul didn't know was that other people were already

thinking the same thing. In response to the 1992 European Habitats Directive, which urged member states to consider restoration of species and habitats, Scottish Natural Heritage (SNH) had already commissioned research into the beaver and were about to launch a consultation looking at the prospect of their reintroduction.

'You can imagine that I was really excited. I was invited to the launch and came away thinking how marvellous, and that these creatures would be back soon,' Paul said. 'But over time it became apparent that nothing was happening. There was talk, lots of talk; there were meetings and reports; but no beavers. So I did a little digging and found that powerful forces were dead set against anything happening.'

But what objections could there be to the reintroduction of the beaver? I wondered. It's not as if they posed a threat to anyone.

'Well, you wouldn't think so, but there was some real hostility,' he said sadly. 'For example, from the Scottish Landowners' Federation, who can be a very vocal group – of which I am still a member, I should add. Then there was the National Farmers' Union, who presented a good case, to be honest.'

Beavers, it seemed, are not necessarily the neighbours of choice of those who farm river floodplains, because beavers can leave holes in river banks that protect valuable crops and treasured homes. Most surprising to me were the objections from fishermen. 'But beavers don't eat fish!' I exclaimed.

'They said they'd build dams that would prevent salmon making it to their spawning grounds,' Paul said with a heavy sigh. 'I just wonder how the salmon coped before we did away with the beavers 400 years ago.'

Moving on up the stream, I called Paul's attention to a felled tree. It was only small, about 7 cm in diameter. It looked as if a few strokes of a billhook had brought it down. Was this Paul's work, or had a beaver felled it?

'Oh, that . . . yes, that's beaver, but just you wait and see what's up ahead.' Paul was nonplussed, but this was the first beaver-felled tree I had seen, and I grabbed a photograph while he wandered on ahead. He had stopped by a small dam. I'd never seen beaver engineering before, either, and if the felled tree had been impressive, this really made me understand the landscape-altering capacity of these rodent lumberjacks.

The dam itself was not very high, about a metre. The stream we had been walking up trickled through its timber wall. But it was when I saw what the dam was holding back that I understood the changes that can be wrought by the animal's presence. Where once there had been a canalised ditch there was now around an acre of wetland.

'This area is my first real success,' Paul said. 'Back in 1997 I entered an agri-environment scheme in which I agreed to allow this patch to revert to wetland. The civil servants who administered the scheme returned in 2002 and decided that what I had done was not enough, so I got my handyman to place a load of stones and make a sort of dam. Then, when they arrived later that year, the beavers took over the management.'

But where did they come from?

'Oh, we'll get to that in good time, but first look at this,' he said. 'You can still see a few of the original stones, but the beavers have been layering up mud and branches on top. And you see how the dam continues?' He pointed towards what looked like a path through the beaver-created bog. 'That extends for over a hundred metres to the patch of aspen on the far side. It's all made by the beavers. As the water rose behind the first dam, the lie of the land caused it to spill north. To prevent it finding a course around the dam, the beavers have simply extended the dam, blocking the flow and providing another extremely useful addition to the landscape.'

And with that, Paul set off, walking along the top of the dam.

'In years gone by, in some of the wetter regions of Britain – the Somerset Levels, for instance – beavers were invaluable,' he continued. 'Not only did they make ponds that acted as nurseries for fish; they also provided causeways like this one over wet land.'

'The committee SNH set up to look at the possibility of beaver reintroduction was rather poorly balanced, I feel,' he continued, stopping on an island in the new marsh. 'It seemed to be mostly made up of people hostile to the idea, and the proponents from the Scottish Wildlife Trust were a little half-hearted. So I decided that there ought to be some sort of beaver network that would act to support the proposal. And in the end the main board of the SNH did accept the notion, but action was glacially slow. This was why I decided that there should be a private enterprise to reintroduce the beaver to Scotland.'

As much as I love the idea of equality, of everyone being given the same lot in life to make of as they will, I am also delighted that there are people who have had wonders placed in their laps and are willing to use them creatively. Which is just what Paul set out to do.

But how, exactly, does one go about getting hold of beavers? And from whom do you have to get permission?

'We did check with the official bodies, and, oddly enough, no one knew of any formal permission that was required. So we were able to go ahead and bring in the beavers.'

How? eBay?

'Louise and I ended up driving down to Kent in a white van to pick up a pair,' said Paul. It was such a delightful image. 'It was not straightforward,' he went on. 'It started with the Kent Wildlife Trust wanting to open up their reserve at Ham Fen, which was becoming overgrown with willow. They hit upon the idea of using beavers. I had, by this stage, made some useful contacts in Trondheim, who were able to get them some from Norway. But when they arrived, Defra kicked up an awful fuss and it was only after the trust managed to get a meeting with the relevant minister at the time, Michael Meacher, that they were able to secure permission. But this resulted in the beavers being in quarantine for nine months rather than the anticipated six. And there had started to be some problems with parasites and disease, so it was far from ideal.'

And that was the start of it all. In 2002 a pair of beavers was released into a fenced section of Paul's estate. Did he know what would happen?

'This is what has been so fascinating,' he said. 'It's an experiment. Even now, nine years on, it's uncertain what will happen. But I'm learning all the time, and I'm sharing what I learn, so that when the whole prospect of reintroduction becomes less contentious, we do them as well as possible.'

A little further on and I stopped us at a substantial ash, a beautiful tree that was over 100 years old. A great bite had been taken from it, and looked to have struck a mortal blow. Was this one of the consequences of beaver reintroduction?

'I think they'll have this one down in time, but the wood is very hard,' Paul said. Did I detect a little resignation in his voice? 'Look at the chips down here.' There were the little gnawings – wood shavings, really. The smaller the chips, Paul explained, the harder the wood.

'But this is nothing compared to the next one,' he said. 'Come on.'

Ahead of us stood a large beech tree, maybe a couple of hundred years old. It was proof that the beavers we know from cartoons are not so far removed from real ones. The tree had a 'waspish' waist, where it had been nibbled to a narrow core. All the way around, the beaver's teeth had chiselled. Lifting themselves up onto their back legs and tilting their head to one side, they gnaw with their long, large, curved incisors.

Each spring I collect and split wood for the following winter. I know the impact this has on the blade of my axe. So how does a beaver manage to keep its teeth sharp?

'I've watched them at work through my telescope,' said Paul. 'Every ten minutes or so they'll stop gnawing away at the tree but continue to chew on nothing. What I was witnessing was the beaver's self-sharpening teeth in action. The beaver uses the hard enamel of the lower incisors to shave away the softer dentine of the upper teeth; then it repeats the exercise on the lower teeth.'

The incisors, as in all rodents, continue to grow during the life of the animal, so can be continually sharpened. The grinding molars, on the other hand, are more vulnerable, and gradually wear away to little or nothing as the beaver ages.

'You know, I never cease to be amazed by these animals,' Paul said, as we walked on along the beaver bridge. Using his walking stick, he was pointing out his planting successes and failures, his attempts to create a good mix of indigenous trees. Then he showed me the beavers' lodge, a mound of branches which housed a wintering family.

The tree felling was certainly not random vandalism but part of a brilliant strategy to help beavers survive in the harshest of conditions. The range of this species spreads across Europe to Russia with its freezing winters, and the close relative of these European beavers in North America has to contend with the icy wastes of Canada. To achieve this, the beaver has evolved into a most 'human' rodent.

Beavers actively and deliberately modify their environment to their benefit. Dam building gives them security as it protects the lodge in which they live and sleep. The lakes they create not only discourage predators but also act as a food store. During the autumn, a cache of food is collected, with the most delectable articles placed deepest in the pile, so that less desirable trunks weigh them down in the refrigerating water. And while this helps keep the food fresh, it also saves it from being locked solid in ice. The lakes also help to boost the growth of the sort of vegetation that the beaver likes to eat, creating coppices, with poplars and willow springing up new shoots from the stumps of the felled trees.

Other construction comes in the form of canals. These give beavers a way of travelling between lakes within the relative safety of the water. It also allows them access to more trees.

'Some interesting research has looked at how much time beavers spend in or near the water,' explained Paul. 'Some populations have been recorded as spending up to 95 per cent of their lives either in the water or within five metres of it. Other populations have generated different figures, but the point is, these are animals that really prefer an aquatic life.'

And by building canals, the beavers open up woodland previously out of reach of their gnawing incisors. They need access to a lot of food because the plants and bark they eat are not very nutritious. Unlike the cow, which can ruminate vegetation with bacteria and draw the maximum nutrition from it, the beaver has to do as the rabbit does, which, in effect, is to double the length of its intestine by making sure the food passes through twice.

'They do eat an extraordinary variety of species,' Paul said. 'In the spring and summer it is much more the non-woody leaves, roots, herbs and grasses, but as the year progresses they focus more on trees. They don't tend to feed where they find the food, but take it off to a special site. The French so often seem to find a better word for things, and they describe what we know as the beaver "feeding station" as a "*réfectoire*". I love the idea of these monastic beavers all chewing in silence.

'And I do hope you'll get to see some eating later; they're so dextrous,' he continued. 'You know that we primates are defined by our opposable thumbs? Well, they have an opposable little finger. Quite delightful.'

As we walked around this segment of his estate we came to an area that looked like a battlefield. Specifically, it reminded me of the harrowing images from the trenches of World War

One: broken trees and large puddles. It was easy to see why the critics might object.

'Well, if you are bothered by the aesthetic, then, yes, this is not the land at its best,' Paul said calmly. 'But consider it, for a moment, from an ecological point of view. This was originally pasture, in which I had planted a mixture of goat- and grey willow. It was a bit of an experiment. Now, you see around the stumps of the grey willow? New life; they're coppicing. The goat willow seems less keen on the wet soil, but even the remaining trunks are a wonderful host to a tremendously diverse array of invertebrates. Anyway, up until a few weeks ago, this area was under water. And the roe deer have not been helpful in allowing the shoots to make headway.'

While this patch of land may not have looked as pretty as the more manicured corners of Bamff, its value was clear.

'Beavers can help set off a wonderful thing called a "trophic cascade",' Paul said. 'In Latvia a scheme was set up to conserve their carnivores, in particular the lynx. They found that areas with beavers had more abundant coppice growth, which in turn created browse and shelter for roe deer – which is the favoured food of the lynx.'

We'd been chatting for over an hour when suddenly Paul realised the time. If we were going to see any beavers, we'd need to get to them while there was still light. So with a more determined stride we headed back to the castle. I wondered about those original beavers he had mentioned; were they still here?

'Our first reintroduction was not perfect,' he said. 'Our first female dropped a tree on herself. It's an occupational hazard, and the long quarantine may not have helped. Maybe she was a little out of practice? Or more likely she just did not have sufficient muscle tone. Then we got another female from Kent, but she was very thin. Then our first male died; he had a heavy infestation of liver fluke. We were only able to find two females to replace him, which left us with no males. It was not until autumn 2004 that we were able to get a male, and from that moment on, well, the numbers have been growing.'

So how many were there now?

'To be honest, I'm not really sure,' he admitted. 'But at this time of year there is a lot more activity, because pregnant females will encourage the rest of the family group to leave them in peace.'

While beavers live within a family group, or colony, comprising an adult pair, yearlings and sometimes older offspring, the female becomes increasingly intolerant as her pregnancy advances, causing the rest of the family to vacate the principal burrow for smaller ones nearby.

By the time we got back to the castle, nature's dimmer-switch had started to turn, so we grabbed a telescope and tripod and headed off downstream. It was clear to me that Paul was very happy with the changes to his land, but I was still not sure if this was a universal position. Had the naysayers been won over?

'Our immediate neighbours are absolutely fine, as are most people in Scotland. But the problem is down on the flood-plain,' he said. 'Impacts could be more dramatic than just a challenge to aesthetic sensibilities. Beavers might increase the risk of flooding to some farmland. But – and this is where it all gets a little contentious – I would suggest that there should be a change in floodplain farming.'

It turns out that Paul and I share a fondness for the work of University of Essex professor, Jules Pretty. He has long challenged conservative thinking on our relationship with the land, and has pointed out that the farmers on the flood-plains of the River Tay were benefiting enormously from crop yields while many others paid the cost in terms of flooding. The attitude on the plain is that, when floods occur, the purpose of the river is to export the water as fast as possible downstream. Artificial flood banks ensure that the fields do not fulfil their traditional function of holding back a portion of the water, which instead steamrollers downstream onto settlements like Perth, where the costs are borne by those whose houses are inundated.

'Beavers will burrow into the flood bank,' Paul said. 'And when the river is in spate the water will swirl into the vacated burrow, making the holes bigger and possibly undermining the bank. But because the plains are so heavily farmed, right down to the water's edge, and there has been little planting of the riverbank trees that help to absorb the power of the water, erosion has become an issue.'

What was more, the very reason the plains were histori-cally so fertile is because they had been used as floodplains. Over thousands of years, mineral-rich silt had been deposited by winter floods.

'I heard of the most outrageous piece of vandalism recently,' Paul said. 'A farmer down on the floodplain was so incensed by reports of beavers on the loose that he chopped down the trees along the riverbank. So stupid.'

So had Paul's beavers escaped from their enclosures into the surrounding countryside?

'Not as far as I'm aware. I'm not the only person who's released beavers in enclosures around here, you know,' he said. 'There are at least another two ventures in the region, and there were reports of beavers living wild before any of these projects started. Now look, we'd better be quiet if we want see anything.'

I wondered if he was avoiding the subject of escaped beavers, but we carried on down his rutted drive in companionable silence. I followed him off the drive into the darkening green, flinching as I cracked a twig underfoot. Soon there was a mattress of pine needles to dampen our steps. To our left there was a slight upwards slope, dark in evergreen, erasing all silhouettes; to our right a lake. Paul gently placed the tripod down and started to scan the scene with the telescope. Quickly he beckoned to me, his grin standing out in the gloaming. I lifted my binoculars and could see what was making him so happy: about fifty metres away, on the edge of the lake, was a beaver, grooming.

The view through his telescope made it clear why he had lugged it with him this far. I could see the beaver in extraordinary detail, right down to the expression on the beaver's face as it ran its claws through its coat. Sitting upright on its hind legs, its flat, hairless tail acting as a support, the beaver had the unmistakable, shut-eyed, blissed-out look that denoted pleasure.

Grooming is vital for the beaver; the fur must be kept clean in order to keep it waterproof and insulating. What I was witnessing, Paul explained, was part of beaver ritual. Every evening, after its first swim, a beaver will attend to its toilette. Starting with its nose and working across stomach, legs and back, it teases the hairs apart using a cunning little comb. The second digits on its hind feet have a special double claw which not only acts to style the beaver's pelt, but has evolved to just the right size for tick removal.

For me, though, the most extraordinary thing was that look of pleasure on its face. It looked like I feel when someone runs their nails across my scalp. Afterwards I asked Paul whether it was a male or female and he pointed out that telling one from the other was rather tricky. They have a cloaca, a single, sphincter-controlled opening, that hides away most of the clues. Apparently it is possible to either use x-rays or gentle palpation of the region to detect the penis bone, but this requires a remarkably tolerant beaver. I thought this one was female, and pregnant, but had no real evidence to base this on. She just looked pear-shaped and happy.

While I had been absorbed in this vision of furry fun, I had become aware of a noise. It was definitely the sound of something chewing. Paul had heard it, too, and was using the binoculars to scan the lake. He reached over and tugged gently on my shirt, pointing to just behind a tree. And there was a beaver, sitting in the shallows, munching vegetation with extreme concentration and vigour like a mammalian mincing machine. Paul moved away to scan an adjacent lake when I spotted a third beaver leaving a tell-tale V as it swam through the water. As I looked up for Paul, I became aware of what was going on around us. The ducks were beginning to settle for the night; and the water, mirror-still now that the beaver had passed, was reflecting the trees on the far bank. Every now and then the growing tranquillity would be disturbed by the cawing of a rook or by a pheasant flapping its wings and sounding like a a starter-motor. A pair of tawny owls were chatting – screech and hoot – from either side of the wood. And, in the twig-fringed patches of sky, bats were flitting back and forth. Beavers, it occurred to me, are good for bats, which hunt insects that thrive in wetland.

Everything about this moment was perfect. The air was calm, warm and resinous. Life was all around us, some of it gearing up for the night, some of it preparing for sleep. It was exhilarating. Misty-eyed as I was, I hadn't noticed Paul beckoning from ahead. His telescope was pointing to the second lake and he looked both excited and a little alarmed. He whispered into my ear, 'There are five.'

In the end I only saw four of them, but this was obviously a great thrill for both of us as we took our leave of the scene. On our way back to the house I told Paul that the experience had been as wonder-filled as anything that had happened to me in East Africa. I felt filled up with delight, in much the same way as I do on leaving a particularly powerful musical performance.

We both grinned our way back to the castle and, as we entered the book-lined hall, Paul admitted that he had been a little surprised to see quite so many beavers in one place. 'In fact, I'm a little worried. To have eight so close together at this time of year means that quite a bit of dispersal is about to happen. I must go and check the fences. But not now; let's go and cook some dinner.'

The night was long and fun-filled. Paul's daughter, Sophie, was visiting, and after dinner she took up her guitar and sang. Rarely have I heard a more beautiful voice, as she caressed old Gaelic songs into life. My room was on the third floor, the old nursery, and I collapsed into bed, smiling like the beaver I had seen grooming itself earlier.

The next morning, on the way down to breakfast, I noticed a pelt hanging over the banister; the large paddle tail connected to it left me in no doubt that it was a beaver pelt. As I stroked it, I could detect two distinct layers of fur: longer guard hairs and a dense under-fur.

Paul made a vat of strong coffee and we talked about the pelt. It had not been one of his beavers; rather it was a gift. 'You felt how rich it was?' he asked. 'Well, that was the beaver's downfall as a species. And it was not just the fur's insulating qualities that were attractive. The under-fur is perfect for felting. And as the beaver was considered such an intelligent animal, well, through the simple application of the principles of sympathetic magic, that makes it doubly perfect for hat making.'

(A good way to get your beaver felt was through recycling. Coats made of beaver fur would be recycled into felt hats, after the long hairs had worn away.)

The cost to the beavers of this desire for their fur was predictable. The European beaver was last seen in Britain in the wild in the sixteenth century and was nearly hunted to extinction across Europe. The North American beaver is credited with stimulating the exploration of the continent's interior in the quest for furry gold. Worse still for the animal, castoreum, the secretions from their glands, known as castor sacs, were considered to have medicinal properties. Indeed, in areas where beavers ate a lot of willow, castoreum does contain high levels of acetylsalicylic acid, or what we would now call aspirin.

The morning was grey but dry, and Paul wanted to show me some more dams before I began the long journey south. I could tell he was quite proud of them, as we strode back down the drive. He talked about the engineering skills of these rodents in an almost proprietary manner. 'There was a bit of a fuss at a beaver convention I went to,' he said. 'The North American beaver people were claiming that their beavers built bigger dams. Okay, so the biggest one ever seen was in Northern Alberta, at 850 metres long, but on the whole the structures they build are of a pretty similar size.'

The dam Paul took me to, while not quite on that scale, was still impressive. It took a bit of getting to, and while I made my way across a slippery log, Paul (of a more mature vintage) had already skipped up onto the top of the dam. This gave me an indication of its scale. It was a good six feet tall. And while the dam I had seen the previous night had created a wetland, this one had created a lake. Quite a big lake – almost an acre; so this construction made by rodents was holding back goodness knew how many tons of water.

And this brought me back to the feasibility of Paul's plans. These animals were notorious for their landscape-changing habits. The new lake on his land was a wonderful habitat for fish, birds and bats. But this is deemed acceptable only because the field into which the lake has encroached is his. How would a similar scenario be received in the big wild world?

'Yes, they make changes,' Paul said. 'But so do we. In fact, nothing has changed the face of the country more than people. We have to learn to share a little. And we have to learn to think around this particular subject a little more. For example, I was called up in 2006 by the local wildlife-crime officer, who asked me if I had lost a beaver. A fishery had just found that its trees were being felled not by alcohol-fuelled vandals but by a beaver. It wasn't one of mine, but I went to take a look and left the fisheries manager with a set of night-vision equipment. He saw two, probably a pair. I then lent him a trap and told him where he should put it. A week later, I found that they had become so enamoured by their new residents that they were considering setting up a beaver-watching business instead.'

And, as last night had proved, to me at least, the beaver offered a wonderful wildlife experience. But the fishery's plan failed, as Scottish Natural Heritage brandished section 14.1 of the Wildlife and Countryside Act and told the owners that if any of the animals escaped they would be liable to an unlimited fine and up to two years in prison.

'So does that mean you are liable to such consequences, if your beavers escape?' I asked Paul.

'Oh, yes,' he said. 'In fact, last autumn I had a visit from a policeman who officially cautioned me. But it would be complicated if it ever went to court. You see, there's some conflicting legislation. European law protects beavers. If they've re-established themselves in an area that's considered to be part of their original natural range they are protected. British law really only deals with individuals, escapees. So if there are beavers out there breeding, then they are protected. This is why SNH is so keen to argue that the beavers they are trying to round up from the River Tay are all individuals. Well, I don't believe that for a moment, and if it takes a court case to decide, so be it.'

I would not want to be up against Paul. He's determined and enormously well read. He's started his own campaign to protect the beavers that are out there already. His vitriol, though, is not aimed at the scientists at SNH. 'Public-relations departments are the real shits-brigade in all this,' he said. 'They see a conflict as something to win rather than resolve, and so they just fight, until they find a better-paid job somewhere else.'

So what does the future hold for Paul's beavers? Well, he's not alone. Aside from the releases in Scotland, and an SNH-sponsored study in Knapdale, there is the project in Kent, at the Ham Fen. These are clever animals, so I don't imagine it will be too long before they've found a way out of their enclosures and into the wild. Then all it will take is a little intelligence on the part of the people who share the environment with them to make sure that everyone benefits.

On the train home I kept drifting back to my night out with the beavers. What I had felt at the time still resonated; this had been as magical an animal encounter as any I had experienced. Could I have a beaver tattoo? They met many of my fairly arbitrary requirements, and the experience had been deeply special. But they are not an animal to which it is easy to get close.

At least that's what I'd thought, but when I went through my notes I found that Paul had answered that already. 'I've avoided being friendly with them because I'm too aware that familiarity with humans could lead to death for a beaver,' he said. 'There are many stories of people befriending beavers in America; there's a lovely photograph I've seen of a woman swimming with a beaver. And I'm sure that ours could also become tolerant of people. But I don't want to risk that, however amazing it would be to get so close.'

Somehow, I thought, this distance was important.

15. ADDERS

*'A wonderful bundle of jumbled
snakes, basking in the sun'*

My final interview had been recorded and I was making the
long journey back home from Alyth and Paul's beavers. On
the train I took stock. I ran through what I had accomplished
and was feeling gently smug. Obviously I was not going to be
able to meet every animal and its obsessive advocate. Britain,
I think, is particular in the extent of its eccentricity.

I wrote a list and, around Carlisle, had one of those
moments of chill in my stomach as I noticed a dramatic
absence. Mammals, birds, insects and amphibians, all covered;
but reptiles? How could I forget reptiles? These fascinating
descendants of dinosaurs are an essential component of the
natural diversity of our country. Damn it, a friend of mine
even had a sand lizard tattooed on her right shoulder at the
ExtInked show.

My list also revealed another startling absence. One particular region of Britain is famed for both eccentricity and wildlife. East Anglia. And yet I had somehow failed to find an
animal advocate there.

Luckily the fantastic Amphibian and Reptile Conservation
(ARC) organisation were not overly alarmed by my request

to find a reptile person in East Anglia. The mainstream names were too busy, but another name kept appearing, always with a caveat. 'You could try Bernard Dawson,' I was told, 'but I don't think he likes people very much.' I was also warned that getting hold of him could be difficult. He didn't have the Internet or an answer-phone, and was frequently out.

By now I had an image of a long-bearded hermit living in the middle of a woodland glade, wreathed in smoke from open fires, grumbling about technological progress. When I did get in touch with Bernard he was polite but reserved, and asked that I write him an email explaining what I wanted. (He goes to the library each week to check up on his electronic communication.) His response was anything but reserved and he invited me to visit him as soon as possible. The best time of year for seeing his totem animal – the species he had dedicated his life to – had already passed. The last remaining dragons, he called them.

I was off in search of an animal desperately in need of a PR makeover. The adder is disadvantaged from the start: it is a snake and it is also venomous. In fact, it is Britain's only venomous snake; indeed, it is Britain's only venomous verte-brate – if you discount the bite of the pygmy shrew.

What can Bernard do to teach us to stop worrying and love the adder? I arrived on the dot of 8 a.m. outside Bernard's home in Holt, a pretty little town in Norfolk. No hermit in the woods, then; his house was in the middle of a fairly modern estate, and everything looked normal. He was stand-ing outside the front door, waiting for me.

'You're on time,' he said, and I got the feeling I'd passed an important test. 'Okay, let's go and find some adders.' He hovered around me as I got into walking boots and prepared my cameras and recording gear. 'I don't think I'm going to like that very much,' he said, pointing at my gin-bottle-shaped digital recorder. I was beginning to see that this might not be the easiest interview to undertake.

'It is about twenty minutes' walk to the reserve,' he said, before setting off with long strides, leaving me scampering to catch up. He moderated his pace as I pulled a hat out to keep the sun from my eyes. 'I have to wear one all the time now,' he said. 'Apparently, years of field work in the glare gave me the cataracts. And I had to get a haircut, too; the hair flicking in my eyes irritated things further. I left the pony tail, though; I don't want to be thought of as normal.'

There was no real danger of that. Bernard was in his sixties, retired since forty-five, and every day he indulged his passion with an energy that left me, as has been the case with many of my ambassadors, struggling to keep up.

'We're heading for Holt Lowes, an area of heathland that is in the process of being seriously trashed by lowland-heath restoration,' he said, when I had caught my breath. He was clearly angry, and as we waited for another large lorry to pass us by, kicking up clouds of dust, he explained. 'They've received half a million quid to do this. These lorries are shifting the topsoil they've scraped off the heath to remove the encroaching scrub – the bracken and the leggy European gorse. What we want here is the *western* gorse; a much smaller species, with a shorter flowering season. The problem is that, while the intention is good – to stop this place turning from heath to scrub to woodland – they're doing it in such a cack-handed manner that they're seriously affecting my adders.'

Bernard was quickly warming up and, far from being put off by my recording him, seemed keen to take control of the interview. 'What I'll do is give you a brief bit about me,' he said, as we continued on our way. I had explained in my email that I was keen to understand the motivation of my ambassadors, and that this meant learning about their own history.

'I was raised in a fairly rural part of Hampshire. In those days you could just let kids wander off, and I very quickly got into wildlife and I used to collect reptiles and amphibians and bring them home. I remember my mother not being impressed when a lizard gave birth in the kitchen. While I was out, I learned to look under sheets of corrugated iron, as it was there that I sometimes found the adders. Every time I saw an adder I got this thrill of total fear – shivers would go up my spine, and I would stare at it, thinking that this creature could kill me. So that was the origination of my passion, it came from the adrenaline rush of being confronted by my own mortality.'

Rather like bungee jumpers and mountaineers, I suggested.

'I was nine and being utterly stupid,' he said. 'The number of herpetologists that get bitten every year as they monitor smooth-snake populations is frightening. They find the snakes by lifting up sheets of discarded corrugated iron, just like I used to do. In fact, I still do it, but I'm a little more cautious these days.'

So had he ever been bitten?

He laughed. 'After all these years, my first bite was last year, in November. I was out taping off areas of heath where I knew adders basked, to try and stop the diggers destroying them. Anyway, I didn't notice it, at the time. I just started to feel a burning sensation on my calf and I didn't pay much attention, because I'm always getting bitten and stung by bugs out here. After about ten minutes, the burning was getting quite fierce, so I had a look, and there was this big red patch and two fang marks. Amazing thing, as soon as I *knew* I'd been bitten by an adder, I started to react. I started to panic, my heart began to race. A guy from the Norfolk Wildlife Trust was there and he took me to the medical centre, but they weren't interested – they told me to go to Cromer, where the anti-venom was stored. And I just thought sod it, and went home. I'd calmed down and decided that if things got bad I'd do a 999 and get help sharpish.'

Just then, we reached the edge of the nature reserve, and sure enough a sign warned of the risk of adders. So how dangerous are they?

'Put it this way,' Bernard said. 'Since records began in 1876 there have been just fourteen deaths. And about a hundred people are bitten each year. Most people react like I did; it was certainly unpleasant but I didn't need hospital. If people do react badly then there is the anti-venom.'

The demographic for adder bites is pretty clear cut. There are the herpetologists, who put themselves in harm's way; then there are toddlers, who'll see a flower or a leaf and reach to pick it, not noticing the snake; and then there are those adolescent boys who consider themselves immortal and are after a thrill – rather like the younger version of my guide, in fact.

Walking onto the reserve, it looked rather like a bomb-site – bare soil flattened by heavy trucks. This didn't seem like a promising place for a wildlife expedition. Bernard, meanwhile, was just about controlling a snarl. I remembered that my bee man, Ivan, had been excited by the soil scraping he had engineered: it removed the bracken and the seed bank in the topsoil, opening up space for bees to mine. But that was a small patch; here it was a vast area.

Further in, Bernard started to point out pockets of vegetation, and when we were through this flattened area, the majority of the fifty hectares still possessed its topsoil. But these pockets of heathland were vitally important. 'Over the last ten years,' Bernard explained, 'I've been mapping the basking areas; these are vital for the adder, especially in the spring. When the air temperature is low, adders use patches of ground that act as suntraps to raise their own temperatures to a level that allows them to function properly. These are usually located quite close to where their hibernacula are – the places where they hibernate. I can't find exactly where the snakes are over-wintering, but by spending every morning walking around here and mapping the basking sites, I've been able to persuade the reserve management to leave these twenty-five-metre-wide patches, which should include the hibernacula.'

But it has not all been successful. Within some of the taped-off areas, workers felled trees without a bulldozer, in an attempt to be sensitive to the adders. But at many of the sites where this happened, the adders had not emerged this spring, and Bernard thinks this is because the entrances to the tunnels in which the snakes were hibernating were squashed shut.

'Snakes don't dig,' he said. 'They use an old mouse hole, or the cavity left where a tree root has rotted away. If they wake up in the spring and find the entrance blocked, they'll simply die.'

It was clear from the way that Bernard spoke about the adders that he cared for them as intensely as I did for hedgehogs; which was strange. There was nothing particularly attractive about them; no history of human affection, as there is with hedgehogs. And I'm pretty sure this is not an animal that you would want to get nose-to-nose with.

'Ha! No, I wouldn't want to get nose-to-nose with one of these beauties,' Bernard said. 'But just you wait until you see one, then I hope you'll understand why they work such magic. I should warn you, the chances are not high. If you'd called me a month ago we might have had more luck.'

Why? The weather was mild and they were obviously out of hibernation.

'Right, you see these basking sites I mentioned?' he said, pointing to the patches of gorse and bracken that had escaped the blade of the bulldozer. 'They're at their busiest in March and April, and that is when I get my highest counts. They come out of hibernation around February – the males first – but are pretty sluggish. The numbers of snakes at each of the forty-or-so sites around Holt Lowes gradually increases as spring approaches, and they'll bask together, wrapped around each other, absorbing the sun. In March the sloughing begins.'

Sloughing – pronounced 'sluffing' – is the process by which snakes cast off their skin, and is a vital part of their lifecycle. 'You can always tell when one is about to slough,' Bernard said. 'The scales that cover the eyes turn milky, because a chemical is secreted under the old skin that separates it from the new, and this includes the eyes.

When the males emerge with their fresh new skin, they are

just so beautiful.' Bernard had gone quite misty-eyed himself. 'You see a bunch of them together, all different colours, they look like jewels. It's one of the wonders of the natural world.'

Adders are sexually 'dimorphic' – in other words, the males and females look very different. 'The females are usually buffy brown, with a distinct darker-brown zigzag along the back,' he said. 'Whereas in the male the background colour is more variable, from almost silver through various shades of green and blue to black. But the markings are jet black. In all but the darkest, this makes the markings stand out far more than in the female. It's when the males are out and about in their new skin, looking for females, that you get a chance to see what I think is the most magical behaviour. They dance.'

'So is there a chance that we might see this?' I asked.

'None at all,' he replied. 'I already told you, you're too late. The dance is part of the courtship, and I'll tell you about that in a moment. After mating the males tend to head off to better hunting grounds, and because the air temperature is warmer and they need to bask less, you don't get to see them. And any females that aren't pregnant will join them. What we are left with now are the pregnant females, and that's what we might just have a chance of seeing, if we're lucky.'

So they stop hunting during pregnancy?

'They don't eat until after giving birth,' Bernard said. 'You have to stop thinking like a mammal. These ancient animals eat and digest in a different way to us – much more slowly. So while the others are off hunting small mammals and lizards, the ones that are left are just gently metabolising the fat from their last meal, which was probably in October. In fact, there is a fascinating way to find out what they have been eating.

Before they settle down for the winter, they'll sometimes regurgitate the undigested remains of their last meal; that gives me a chance to have a look. Around here they eat a lot more frogs than in other areas, judging by what I find. They have to get the remains out of their system because the active digestion pretty much stops while they hibernate, and the last bits of frog would just rot.'

'And I thought *my* morning breath was dragon-like,' I joked.

'So you know about the dragons?' Bernard asked, as we walked on through the gorse.

I was lost. What dragons? I was just talking about fiery breath.

'Almost as much as adders, I love dragons,' he explained. 'I think the adders are the last of the dragons. They might not breathe fire, but they have venom. The legends of fire-breathing come from the snake's forked-tongue, flicking out to sense the air. That's how they smell. They pick up particles of the air on the tips of their tongue, and pull them into the roof of their mouth, where there's an organ called the Jacobson's organ. This process is sort of like smelling, but different. It bypasses what we think of as smell, sending messages from the air straight to other parts of the snake's brain. It's more like the way pheromones work in mammals.'

(Later, back at his house for lunch, Bernard's fascination with dragons became evident. Most of the rooms were filled with dragons: paintings and ornaments and even some delightfully fanciful 'skulls' sitting alongside real fossils in a cabinet in the dining room.)

Now, on the heath, we had stopped by another patch of

scrub. I'd already accepted this was going to resemble the experience I'd had with the moths, and wasn't expecting to see anything. 'Thing is,' Bernard said, 'that tree stump there used to have scrub all the way around it. It created a little microclimate, shielded from wind but catching the early sun. I have a picture of ten adders sitting at the base of the stump on the south side. But now they've removed the shielding vegetation it's as cold as anywhere else, and there've been no adders. The guidebooks will tell you that adders don't come out until the air temperature reaches ten degrees centigrade, and that might be true. But that figure refers to the shade temperature, measured at a single spot. I bet if you were to put a thermometer down there, back when it was a sun trap, the reading would be much higher. I have photographs of an adder basking next to a patch of snow. And you know they don't have eyelids, right? Well, that means that when it starts to get too bright they have to hide away. Which means that if you wait until the air temperature is ten degrees, well, you'll be too late for most of them.'

He led the way to another adder patch, this time through knee-high heather. 'In the spring, when the adders are most visible, I always wear Wellington boots. When that one got me last November, I really didn't think there would be any around. And now, well, like I said, most are gone.'

Even so, I was finding it difficult to concentrate on operating the audio recorder while scanning the ground ahead for signs of danger.

'Don't worry. They'll scarper, most likely, with us making this racket. They don't have ears, but they can sense vibrations, and we're making plenty of those. And they see movement too, and they smell us. They're not aggressive; they'll tend to flee as a first line of defence. Judging by where the fang marks of the one that bit me were, on the back of my calf, I must have backed into it while I was working. It would have performed its usual threatening behaviour – a little hiss, then they rear back their head. That's when you should move away – but I didn't see it, so it got me.'

Gingerly I continued, and we came out on a grass path. 'Up here there's a good basking spot and the wind is blowing towards us, so let's keep quiet for a bit.'

We crept closer. I stopped beside Bernard and we both stared into a miniature clearing. 'I keep that one trimmed,' he whispered. Nothing. But as Bernard turned to move on, he hissed and pointed, just in time for me to see the rear end of an adder disappearing into the bracken.

'Did you see that?' he said, grinning. He looked as thrilled as I felt, even though she'd looked pretty small. 'It wasn't that small. Perhaps you only saw a little part of it,' he said, a little defensively. 'To be here, now, she had to be pregnant, which means she will be at least three years old, so probably nineteen to twenty inches long. Anyway, that was really encouraging; it means there are at least some around.'

It would be easy to walk right by an adder without noticing it. The adder's pattern was beautiful, but very similar to the dried bracken on which she had sat. Only Bernard, with his eyes attuned by many years of watching, had allowed me to see her.

As we walked, Bernard talked more about the impact on his adders of the lowland-heath restoration efforts. Habitat management is never going to please everyone. On Holt

Lowes an attempt was being made to prevent the ecosystem's natural progression to woodland. One of the things we love about nature is its fluidity; and yet often we have to fight against it. Given enough space, as Holt Lowes evolved into woodland, other areas would become heath, due to fire or other ecological shifts. But there is not enough space, so we manage what we have and attempt to stem the tide of change.

'I reckon this work has destroyed maybe 25 per cent of the adder population,' he said. 'But if I hadn't been mapping the basking areas, and if the Norfolk Wildlife Trust hadn't helped me tape them off from the bulldozers, then the hibernacula of many more would have been crushed. I reckon we would have lost at least 80 per cent of the adders. In the end, yes, the habitat will be better for adders. But for now I'm not happy. Just look at that slope.'

The area he was pointing at was bare earth, where once it had been scrub. There were no hibernacula that he knew of here, but the expanse of flat soil had created another problem. 'Up there,' he said, pointing to a taped-off patch of gorse and bracken, 'is a fantastic basking site. I know that males will leave that spot to go hunting down on the lower ground, there where it is a bit boggier. They used to be able to make that journey under cover of vegetation, but now there is no vegetation, and the snakes will be vulnerable to predators. I dread to think how many will have been killed and eaten.'

I tended to think of adders as predators, not prey. What, I wondered, were their main threats? 'Buzzards and other birds of prey are the traditional predators, but what gets me angry are the pheasants,' he said. 'They're an alien species, bred to be hunted, for the pleasure of a few, and they are ecologically disastrous. They'll work their way down a hedgerow and take absolutely everything, all the insects, amphibians and reptiles, even young adders.'

As we followed a path down a slight slope, Bernard pointed to the raised bank on our left. 'Dog walkers come down here every day,' he said. 'And they walk past maybe a dozen adders, just basking on that south-facing slope, and they never see a thing. Mind you, some of the dogs do, and they tend to come off rather badly when they go sniffing in the undergrowth.'

We got to another basking area and this time something was there: a skin, the exuvia of an adder. As I reached forward to pick it up, Bernard grabbed my arm. There was an adder very close to the sloughed skin – my first full sighting , even if our movements quickly scared her off.

'I thought you might like a skin,' Bernard said. 'So I have a couple back at home for you. She probably wouldn't have gone for you, but it's better to be safe.'

My mouth was dry; the sudden surge of adrenaline had perked me up, too. We are instinctively scared of snakes, a result of 200,000 years of evolution. This fear, known as biophobia, is deep-seated. In our ancient past we would have learned to avoid things that were dangerous, and that learning has stayed with us. The inverse of this has been the instinct to love certain aspects of the natural world – biophilia, which I believe is as deep-seated as our instinctive fear of snakes, and just as important. We *need* wild experiences.

Beginning to calm down after my encounter, I followed Bernard as he continued to walk. I wondered whether it was just adders, or had his time out here given him a love for other wildlife as well?

'Of course, everything is interconnected, but one of the special things is nightjars,' he said. And what amazing birds they are, with such moth-like plumage and a diet rich in moths. Normally you only get to hear the churring song of the male. 'Only the other day I was looking at an adder, just over there, in fact, using my binoculars. I was about to walk away when I thought I'd have just one more look and noticed, just two feet from the adder, a nightjar, utterly brilliantly blended into the vegetation. The other animals I am rather fond of are dragonflies.' Well obviously, given his dragon fetish. 'It's not just the name. In the summer the adders are not out basking and much harder to see, so I come here and help with the dragonfly surveys.'

I reminded him about the adder-dancing he'd promised to tell me about, and he said he would take me to the spot where he had his best sighting of this behaviour. As we neared a fence, he pointed to the bank on which it was built. 'If you speak to the old guys,' he said, 'they'll tell you about this boundary bank. This is where they would come with air rifles and shoot adders. Now adders are protected, but back then there was a huge amount of prejudice. This bank is interesting. Forty to fifty years ago there were enough adders here to make a shooting expedition worthwhile, but since I've been coming here I haven't seen a single one. The only reason I can think of for that is that conifer plantation,' he said, pointing out about half a hectare of Norway spruce nearby. 'It grew and shaded the bank, so it's no longer a sun trap. But last year they cut a whole chunk of the plantation and with the returning sun came adders, I found seven.'

We went to look at the bank. 'This is the site of the best

dance I have ever seen. When a female is coming into season she attracts the attention of males. One will hang around, sort of guarding her, until the time is right, and others might attempt to take his place. Usually there is a size difference, and the smaller male backs down. But if they are roughly the same size it can get really interesting. What was so great when I saw the snakes dancing here is that they were up on the bank, pretty much at eye-level, so I got a great view. First they reared up, opposite each other, then they moved together. It really did look like they were dancing. Then they coiled around each other, and attempted to force each other to the ground to show who was the stronger; this was followed by much thrashing from side to side, and chasing through the undergrowth. It probably only lasted for a couple of minutes, but it was one of my all-time wildlife highlights.'

As we walked further along the bank, there was another arm-grabbing moment. There, sitting still in the sun, was a female adder. This time, again with the wind in our faces, I had time to pull out my binoculars, and wallow in the alien beauty of her world. This sight was triggering all sorts of atavistic reactions; the fascination I felt went beyond what I have felt when confronted with more benign species. I was transfixed by her mahogany-red eyes with their dark vertical slits, and her darting black tongue. I could see where the stories of snakes hypnotising their victims came from. There is a terrible charm of potential death.

Okay, I know that is a little dramatic, but this is the only venomous animal I will meet in the wild whose bite could kill me.

The few photos I took reveal how cryptic the adder is, the

zigzag pattern of diamonds on her back strikingly similar to the dried-out bracken around her.

Perhaps there was a reflection of light from my binoculars, or maybe I made more noise than I thought, putting the camera down, but my last view of her was a fairly relaxed departure, and her signature sigmoid seemed to go on for a very long time. Perhaps it was time to respect them a little more.

'They can reach around twenty-three inches while they are breeding,' Bernard said, as we started to breathe more normally again. I hadn't realised quite how much I had been holding my breath during our encounter. 'By the time they reach eighteen or twenty years old—'

I had to interrupt: 'They can live for twenty years?'

'Oh yes, they can reach thirty. Anyway, they stop breeding when they get to around twenty, and more of their energy is put into growth. In fact, they start growing again, and can reach up to thirty-one inches.'

As we started back up the slope, Bernard declared himself very satisfied. 'That was more than I thought you would see,' he said. 'You know, if you'd come at the right time you'd have had a ball. Once I saw 165 adders during a single walk.' Seriously? He must have been tripping over them. 'Oh yes, that's why there are warning signs. But most people won't see a single one. It's just that I know where each of the basking sites is and go straight there. If I move briskly, and in a sensible order, I can be sure that I haven't counted any more than once.'

The morning had gone quickly; already it was time to head back to Dragon Central for some lunch. On the way,

Bernard told me about some research he had been help-ing with. 'All over the country, adders are in trouble,' he said. 'I'm just lucky that this is one of the best places in the country to see them. One of the worries is that the remain-ing pockets of adders are too small to sustain themselves – there is not enough genetic diversity, in other words.' Habitat fragmentation. Everywhere I have been this subject has come up: populations rendered unable to mingle, move and replenish by human interventions in their habitats, and being forced to a point where piecemeal extinction is a real threat.

'The project involves drawing DNA samples from fifteen sites around the country,' he continued, 'and assessing how similar the samples are *within* each site and how different they are *between* sites. The thinking is that if we find, for exam-ple, a site with lots of inbreeding, as would be indicated by very similar samples, we might start introducing snakes from other sites to improve the genetic strength of each popula-tion. They might even bring some in from Scandinavia.'

On the way back, we stopped at another basking site and, after a few moments, I felt the familiar tug on my arm and followed Bernard's outstretched finger. And there was another adder. I lifted my binoculars, and gasped. It was my turn to tug Bernard's sleeve. 'There are two!' I whispered. We were both grinning like kids. He got out his binoculars and we stood there, staring at a wonderful bundle of jumbled snakes: two mothers-to-be, basking both in the sun and – unwittingly – in the adoration of two admirers.

As we left the site I experienced a pang of grief. This had been my final meeting with an animal ambassador, my final excursion into someone else's wild world. I had enjoyed myself so much. I didn't want it to end.

But it wasn't over. I'd forgotten something significant. I still had to make a decision: which of these fifteen amazing animals was going to adorn my leg for the rest of my life?

Conclusion

So which creature had seduced me? Had my eye been caught by the well-turned heel of a badger? Had the fluttering eyelashes of a water vole or the freckly nose of a brown long-eared bat won me away from the charms of the hedgehog?

My first and last tattoo would be a hedgehog – that was the original plan. A nice, neat conclusion to my midlife crisis. I had lived with the hedgehog as my totem animal for a quarter of a century. But I'd made a pledge before starting this enterprise: the animal that most appealed to me would be tattooed on the other side of my leg to the hedgehog.

Before embarking I had asked myself a question. Would I really be able to find an animal that felt as important to me as the hedgehog – one that I would feel happy about sharing the rest of my life with? I honestly thought that the choice would be easy – that I would find another animal with which I felt great kinship, an animal that would prove to be the ideal tool for entertaining, enthusing and educating others about the wonders of wildlife.

In the cold light of completion, however, it was not so simple. This was deeply serious for me. I needed to be absolutely sure that I made the right choice. How to measure? How to score? I had toyed with the idea of a completely subjective mechanism – the RICH-T index: 'Researchability' (how easy it was to research), Importance, Cuteness, 'Helpability' (how much it needed assistance) and quality of the Time spent with the animal. I then gave each animal a score out of ten for each variable. The ranking this generated was as follows, with the scores represented as a percentage of the total available:

Toad	78
Bat	72
Otter	72
Robin	72
Bee	64
Water Vole	64
Beaver	64
Badger	62
Fox	62
Adder	60
Dolphin	60
Owl	56
Sparrow	50
Moth	46
Dragonfly	40

Obviously this is utterly subjective, but I went through the list and scored each animal as honestly as possible. So while on one level it was little more than a parlour game, it did reveal some interesting truths about my experience. For example, my time out with dolphins was amazing and scored 9/10. But they fell down on the cuteness stakes, managing just 2/10.

Owls also came surprisingly low, considering what an amazing encounter I'd had with them. But their overall scarcity did lessen the delight.

Before setting out I had wondered how many of my ambassadors would recognise themselves in Byron's description from 'Childe Harold's Pilgrimage':

> There is a pleasure in the pathless woods,
> There is a rapture on the lonely shore,
> There is society where none intrudes,
> By the deep Sea, and music in its roar:
> I love not Man the less, but Nature more

There are times when I myself feel the appeal of nature's solitude, but I cannot deny my generally gregarious nature. I suffer with a surfeit of empathy for both nature and humanity. Most of the guides I had met seemed remarkably well-balanced; it would hard for a misanthrope to project such enthusiasm.

I doubt the moths would have flown even as high in the rankings had it not been for Amy's passionate energy. Another of the invertebrates had also rather failed to make the grade. Beautiful as a dragonfly tattoo might be, I couldn't make enough of a connection with the alien beast.

I was attracted to the mundane nature of the house sparrow. Not long after seeing Denis, I was walking alongside a privet hedge. It was a still day, yet a portion of it was shaking, as if in a breeze. As I got closer, I could hear the chattering of a complex soap opera of sparrows. I was walking to Oxford's irrepressible Magic Cafe where I was served by a woman with a beautiful sparrow tattoo on her arm. Were these signs? In the end I felt that, individually, these birds lacked a certain something – another reason for the invertebrates' low scores.

Bernard with his snakes had made a great impression upon me, and I had been won over by the beauty of the adder, but as Bernard himself had said, this was not an animal you would want to be nose-to-nose with, and while the opportunity for physical closeness was not a defining essential feature of my quest, it was important. Badgers were at a disadvantage on account of their having killed some of my hedgehog friends, but if I had been able to feel the rubbery nose of an inquisitive individual, and stroke its tough hair, I might have been won over. Witnessing the deep love that Gareth felt for those beasts had been deeply moving.

I did have an ulterior motive for wanting to be seduced by the badger, though. In fact there were three. First, a dear friend's nickname is 'badger'. Secondly, as a springboard for ecological debate, you could not choose a better couple. But the main reason was the joke with which I had hoped to end the book; I could have then said, with a straight face, that I had an asymmetric intraguild predatory relationship between my legs.

The beauty of foxes has always held me, but the greatest lesson I had learned from my time out with the good professor was that *understanding more does not mean appreciating less*. My very first chapter thrust into my mind something that I tried to remember throughout my journey: the importance of close observation. Not just looking, not just *seeing*, but becoming completely aware of what is around you. My encounters with beavers and water voles felt strangely similar in retrospect, partly on account of their passionate and articulate advocates. (Concerned that I'd seen the beavered land at its least attractive, Paul Ramsay has since sent me photographs revealing why he considers this animal to represent such a wonderful contribution to the environment: where I had seen a battlefield there are now fields of rich, lush green.)

And what is it that bats, otters and robins share? A very different mix of scores had given them the same total. One of the big things they had in common was the enthusiasm of their proponents, though Andrew's was a more intellectual and considered enthusiasm than Mim and Huma's visceral response to their animals.

But I don't want the wonderful advocates to feel that a low score reflects a lack of enthusiasm or advocacy on their part. Mim scored extra with the otters because of her persistence and generosity after our meeting. Soon after I returned to Oxford a package arrived in the post. It was a small box, not much bigger than a pack of playing cards, elegantly wrapped in handmade paper decorated with a beautiful sketch of an otter, and topped off with a blue ribbon.

Normally I rip presents open, but this demanded delicacy, not least to save the beautiful sketch that adorned the paper. The box had once held jewellery but now held a small plastic bag padded with cotton wool. At once I had a thought about what might be contained within, but this was contradicted by the thought that Mim wouldn't do that; I could not imagine anyone collecting the poop of a wild animal and packaging it up with such refinement, and then posting it.

But I was right. Sure enough, within the bag was a small portion of spraint. I held it up to my nose and inhaled, deeply;

too deeply, as I had to remove a few bits of dried grass that had escaped into my nose. I tried again, and this time, I got it. There was the mix, and it was as Mim had described it: jasmine, grass, tea and a hint of fish. I was astounded. I had assumed this was a joke that was perpetrated on newcomers to the field that would result in great amusement as the novice gagged on a noseful of skunk-related effluent. But no, this was actually very pleasant-smelling faeces. This made me realise that the otter had done remarkably well at getting under my skin – or perhaps Mim had done the job for it. Here was an animal that had been in my consciousness for a long time – one that was beautiful, mysterious and, though this might be an effect of having spent time in and around Totnes, dare I say it, magical. The intelligence of the otter, its gracefulness and beauty: in a way they counted against it. I didn't want to be swayed by aesthetics alone. Beauty is about more than just fur and dancing eyes.

The bat's success was almost entirely down to the passion of Huma. She had managed to excite me with shadows and echoes, but it was the appearance of the brown long-eared bat that did it for me. Getting so close to such an amazing and alien-seeming creature was unexpectedly affecting. If I had been able to handle the bats myself, that might have swung it completely. And Huma has not let up. Very gently, every now and then, I get another invite to join her on some bat-based adventure. One email invited me to join her deep in the bowels of London, where bats hibernating in a secret tunnel needed counting. Maybe next year. Because I don't imagine that my newfound friendships will wither; neither those with people nor those with animals.

In late February there was a hint of spring in the air. I opened the windows to air the shed in which I write, and was hit by the limpid song of a robin. I was also struck by a thought. Could I follow instructions Andrew had given me for robin seduction? I went into the house and dug four small chunks of mature organic cheddar from the block in the fridge (only the best for my robins).

The robin was preening in the bare dogwood, its breast clearly orange against the claret tangle of twigs. I placed the cheese on a wooden chair and retreated. The robin continued to preen, right wing raised, its feathers pulled through the beak, then the same procedure with the left wing. A great tit landed on the dogwood and seemed to observe the scene, before taking off in pursuit of early emerging bugs. A coal tit flitted by, and still the robin preened. I could see the robin and the robin could see me.

Two crows flew over. I became more aware of the hubbub of suburbia: a siren, a starting car, an ice-cream van taking advantage, like the great tit, of the unseasonable warmth. Then the sound of the van's prey playing happily in the park.

For twenty minutes the robin stayed still and watched me watching it. Then I had to go inside to cook dinner. When I came back out at dusk the robin was gone and the cheese remained untouched. I tried again the next day, and the next. Nothing. Perhaps if I had been more determined and less distracted; if I had been presented with a more accommodating robin, its ranking might have been higher.

Another common theme of the high-scoring animals was the depth of the stories that accompanied them. And there was something else about the people I had met. By focusing on a particular animal they had each learned a great deal more besides. Their animals had acted as a lens through which they could see more clearly the world around them. Gareth and his badgers demonstrated this better than most. It was about more than merely observing; it was about *loving*. Without wishing to question the wisdom of the Ancient Greeks, with their six kinds of love, I do think they missed out one of the most crucial varieties. And that is love for the natural world, for the wilds from which we sprung and of which we are still – though we may fight against the idea – a part.

I am drawn to the concept of 'biophilia', the idea that we have an innate need to be in contact with nature. It strikes me that the word could be used to express the seventh variety of love. Clearly we need those other varieties – erotic love, the love of friends, playful love, pragmatic love, self-love and universal love. But I believe we also need love of nature.

And each of these wonderful guides, through their visions and versions of love, has helped me see more clearly. They have all recognised beauty in beasts of such divergent sizes, shapes, appearances and behaviours.

My challenge has been met, many times over. I accept that there are many other animals that share the charismatic appeal of my hedgehogs. And what I've learned from this is that, by applying ourselves to just one aspect of the beautiful and alarmingly fragile diversity of the natural world, we can learn to love not just a single species, but the entire web of life that sustains it and us.

So much of our awareness of nature is remote and mediated through the television. I'm not suggesting that a dose of David Attenborough is not a good thing; but staying glued to the sofa is not enough. As my ambassadors have shown, the world of nature is easily accessible to us. We are, famously, a nation of animal lovers; but that love is not restricted to pets. Remember the robin? Our relationship with this bird, and the trust it shows us, are unique to Britain. It is the *wild* part of wildlife that I think is really important – being out there, immersing ourselves in it.

There is a growing interest in 're-wilding'. This is usually talked about in terms of returning habitats to an older and richer form. But I think my advocates have shown that we might also try to re-wild *ourselves*. Children don't need re-wilding. They have an innate wildness that society often prefers to see tamed. But taming children means they miss out on something special, the development of their sensitivity to the beauty of the world around them. I believe that the author and zoologist Konrad Lorenz had it right when he wrote that children should have 'the closest possible contact with the living natural world at the earliest possible age'. The natural world, he said, was 'the best school at which young people can learn that the world has significance and meaning'.

The closer we get to nature the better our chances are of finding a 'gateway species', one that can become our personal totem animal. I loved the idea developed by Philip Pullman in the 'His Dark Materials' books of everyone having their own daemon, their own connection to the wild, intimately linking them to something other than just themselves. When we have one powerful connection we find that our entire perspective shifts and it becomes easier to fall in love with nature. And to reiterate the words of Stephen Jay Gould, we will not fight to save what we do not love.

My route to understanding this 'bigger love' was the hedgehog. I realise, now, that to make my work easier I had to understand more, become more aware of what it was like to *be* a hedgehog. Intra-species empathy might seem an odd proposition, but I believe that by getting close to wildlife, either physically or metaphorically, it is possible to start this process.

But back to the ranking. What was it about the toad that had put it at the top? It was not as if the toad were an iconic species in my life. I had not had a 'toad moment' to compare the encounters I had shared with hedgehogs, foxes, owls, bats or even dolphins. I suppose it is the *potential* that the stories about the toad reveal that make it so appealing. Its ability to change, and the cryptic beauty that lies beneath the unbecoming skin, to be revealed only when its jewelled eye gazes into you. I was wary of encounters that were simply about the quick-fix of instant aesthetic appeal. There is a risk of falling into an affair with nature that is purely sentimental if the chosen gatekeeper is too obvious in its appeal.

Sitting around a fire one evening, cuddling an enormous stag-hound called Bear, I was talking with a wonderful story-teller, Chris. He found my attraction for the toad very interesting, especially in the way that it is almost the antithesis of the hedgehog, engraved on my left leg. The hedgehog's spines act to repel; the skin of the toad is one great sensory organ, absorbing messages from the world around. Apparently the skin of the toad is as sensitive as the skin inside our nose. It paints a picture of its surroundings by absorbing information through its skin. The hedgehog repels contact with its skin.

I had gone looking for my toad through dance – perhaps

the oddest of all the journeys I undertook. Given the climatic conditions, it was hardly a surprise that there had been no mystical meeting. But I wanted a meeting with a toad before I made my final decision. I could not very well go and get a tattoo of a toad simply on the warty back of magical tales.

The dance with which I am comfortable, the sweaty excess of 5 Rhythms, leaves me in a state not a million miles away from the trance described to me by Gordon. Whether they are the same state with a different vocabulary, it is impossible to know.

Cycling home from the class through south Oxford in midsummer I described my predicament to the friend I was with. I really needed to have an encounter with a toad, I explained, and the sooner the better.

We were right by a lake in which I knew they bred, and I'd seen a toad along the path before. I told her I was considering cheating a little, and transposing that previous meeting to the present day, in order to give my story a sense of completion. But as I finished the sentence, there, on the path, was a young toad.

Only once before had I been so excited to see a particular animal, and that was when I discovered that George, one of my radio-tracked hedgehogs, was still alive. My companion, on the other hand, was a little bemused; as was the toad.

I picked it up and gazed at it. Beauty is very much in the eye of the beholder, but as I looked I also realised that beauty is also in the 'I' of the beholder; I realised that my journey had changed my perspective. I was looking at this little amphibian in a way that I would not have before my quest began. I was tempted to enact the fairytale and kiss the toad. And it dawned on me that the toad's fairytale transformation into a prince was utterly apt. The idea of beauty being hidden within the beast is not new; it has been a staple of stories for generations. But we have been made to forget the lesson of those stories by the visions of perfection that confront us at every turn. In the media unobtainable beauty is presented as an ideal that can be achieved through consumption. Seeing *true* beauty, on the other hand, requires us to spend a little time and develop greater awareness. We look at life through the equivalent of a low-resolution camera most of the time. It is only when we pause and look more deeply that we truly become aware.

After my hedgehog tattoo had been completed I was asked, as the final part of Jai Redman's work of art, to sit in an uncomfortable chair and look towards an ancient box. It was a 120-year-old plate camera, and each image was taken with an individual plate of film, with an exposure of several seconds. This was as far removed from my digital camera as was possible to get. And it did a very different job. The amount of information it was capturing was enormous, and when the prints were eventually produced they seemed to look far more deeply into the person than my snaps.

Seeing more deeply and falling in love both come with risks. The pain we experience as a result can be as immense as the pleasure. 'The stabbing pain of love's awakening', to borrow a phrase from Mahler's 'Song of the Earth', is joined with the fear of loss. Because with the appreciation of nature comes understanding of the damage that has been done to the natural world. The understanding that there has been a decline in hedgehog numbers of 95 per cent since 1950. We are blinkered to the true state of the demise of the life around us due to 'shifting baselines'. We accept the density of wildlife we see in our youth as our 'baseline' and strive to return it to such a level, a level which has incrementally shifted our perspective on what is reasonable. Our wild world has suffered such destruction it is no wonder we turn to poets like John Clare for solace.

Thanks to the work of their advocates, each of these fifteen species have shown that they can act as a gateway through which we can learn to see more clearly the world around us. It was a night out with Nigel, one of my hedgehogs, that first allowed me this opportunity, changing my perspective on the animals I was studying and, through the moment of getting nose-to-nose with him, seeing a glimpse of wildness and understanding that this is what we need. And this is a way of getting back a connection with the world around us that will enrich our lives. So much of what we, as the animals we clearly are, have lost has been because of the dislocation from the world in which we evolved. We forget that every bit of 'now' is steeped in millions of years of evolution. We need to learn to see the bigger picture, and one way of doing that is to develop a relationship with nature through a gatekeeper species. So whether it be the bee or the toad, the otter or the fox, or any one of the other wild species of Britain, let us all take the risk of becoming attached to something out there, in the wild world. See deeply, open up your senses, become aware of the interconnectedness of life, and risk falling in love.

Epilogue

Ink vs Steel, Leeds

I did not go straight into the tattoo parlour. I lingered outside, talking myself through the decision I was about to take. Was I really going to get another tattoo? I had come so far on the back of the idea that I would, so running away was impossible. But going in felt like an enormous step too. So with a deep breath, I walked purposefully into Ink vs Steel.

I could have had a tattoo done in Oxford, but I wanted the connection with the ExtInked project, it felt like a good resolution. And when I had called up the proprietor, Simon Caves, who had done my hedgehog, he said he would do it for free as well.

The actual image was easy. I had originally thought I should use Jai again, but as soon as the toad appeared at the top of my list I knew I would have to ask Gordon MacLellan, my guide into the world of the toad, if he would complete my journey by designing my very own toad. Which he did, adding a wonderful twist: the word 'beauty' woven into the lumpy back of my second, and last, tattoo.

Acknowledgements

My ambassadors have been amazingly hospitable and generous, giving me vast amounts of their precious time, sharing the deep love and knowledge they have for their animals. They have also had to put up with occasional spasms of questions, as I tried to clarify the finer details of ecology or conservation. I hope they feel the book is a fitting tribute to the animals they hold so dear, and that it contributes to a wider love and appreciation of the amazing beauty lurking behind even the most unexpected skins.

There was a time when I was worried this book would not happen, but my agent, Patrick Walsh, had a great and determined faith, pushing me into being better than I thought I was, and for this, and plenty else, I am very grateful. Not least for introducing me to my editor, Kerri Sharp, at Simon & Schuster.

At the heart of the book is a love of wildlife, and whenever I hit a tricky patch it was to wildness I turned. Stomping among trees, or dancing into a frenzy, I could rely on the wild to bring me back to my senses. Very rarely did this happen alone. I love the fact that I am surrounded and supported by the most amazing community of friends.

At times I felt slightly guilty at having quite so much fun

while working, but it would not have been possible to be away meeting amazing people, locked in my shed writing about them, or hiding in with my mother to polish in peace, without the very tolerant trinity of my wife and children.

Thank you all.

Further information

I have a website and blog on which you will find photographs and podcasts from my quest to find 'The Beauty in the Beast' – www.urchin.info

There are detailed monographs on all the animals I mention in this book. Many can be found in the New Naturalist series published by Collins, though for a gentler start, the Whittet series of wildlife books is also very informative.

And there are also groups who are involved with either the conservation of or research into the species I have been investigating. Here is a non-exhaustive list of groups that I have found particularly useful and helpful:

Hedgehogs
The British Hedgehog Preservation Society
Hedgehog House
Dhustone
Ludlow
Shropshire
SY8 3PL
01584 890801
www.britishhedgehogs.org.uk

Solitary Bees

Bumblebee Conservation Trust
School of Biological & Environmental Sciences
University of Stirling
Stirling
FK9 4LA
www.bumblebeeconservation.org.uk

Badgers

Badger Trust
P.O. Box 708
East Grinstead
RH19 2WN
08458 287878
www.nfbg.org.uk

Bats

Bat Conservation Trust
5th floor
Quadrant House
250 Kennington Lane
London SE11 5RD
0845 1300 228
www.bats.org.uk

Dragonflies

British Dragonfly Society
23 Bowker Way
Whittlesey
Peterborough
PE7 1PY
www.british-dragonflies.org.uk

House Sparrows
British Trust for Ornithology
The Nunnery
Thetford
Norfolk
IP24 2PU
01842 750050
www.bto.org

Water Voles
The Mammal Society
3 The Carronades
New Road
Southampton
Hampshire
SO14 0AA
023 8023 7874
www.mammal.org.uk

Moths
Butterfly Conservation
Manor Yard
East Lulworth
Wareham
Dorset
BH20 5QP
01929 400209
www.butterfly-conservation.org

Dolphins
Whale and Dolphin Conservation Society
Brookfield House
38 St Paul Street
Chippenham
Wiltshire
SN15 1LJ
01249 449500
www.wdcs.org

Cetacean Research & Rescue Unit
P.O. Box 11307
Banff
AB45 3WB
Scotland
01261 851 696
www.crru.org.uk

Owls
Hawk and Owl Trust
P.O. Box 400
Bishops Lydeard
Taunton
TA4 3WH
0844 984 2824
www.hawkandowl.org

Robins
RSPB
The Lodge

Potton Road
Sandy
Bedfordshire
SG19 2DL
01767 680551
www.rspb.org.uk

Foxes
The Mammal Society (see above)

WildCRU
www.wildcru.org

Toads
Froglife
2A Flag Business Exchange
Vicarage Farm Road
Fengate
Peterborough
PE1 5TX
01733 558844
www.froglife.org

Otters
International Otter Survival Fund
7 Black Park
Broadford
Isle of Skye
IV49 9DE
Scotland
01471 822 487
www.otter.org

Beavers
www.scottishwildbeavers.org
www.bamff.co.uk

Adders
Amphibian and Reptile Conservation Trust
655A Christchurch Road
Boscombe
Bournemouth
Dorset
BH1 4AP
01202 391319
www.arc-trust.org

Tattooing and ExtInked
Ultimate Holding Company
www.uhc.org.uk

Ink vs Steel
51–53 Wade Lane
Leeds
West Yorkshire
LS2 8NJ
0113 243 6162
www.inkvssteel.co.uk